Sabrina Ortlepp

Late Quaternary environmental history of Taylor Valley,

Sabrina Ortlepp

Late Quaternary environmental history of Taylor Valley,

southern Victorialand, Antarctica, reconstructed by a multidisciplinary study of lake sediments

Südwestdeutscher Verlag für Hochschulschriften

Impressum/Imprint (nur für Deutschland/ only for Germany)
Bibliografische Information der Deutschen Nationalbibliothek: Die Deutsche Nationalbibliothek verzeichnet diese Publikation in der Deutschen Nationalbibliografie; detaillierte bibliografische Daten sind im Internet über http://dnb.d-nb.de abrufbar.

Alle in diesem Buch genannten Marken und Produktnamen unterliegen warenzeichen-, marken- oder patentrechtlichem Schutz bzw. sind Warenzeichen oder eingetragene Warenzeichen der jeweiligen Inhaber. Die Wiedergabe von Marken, Produktnamen, Gebrauchsnamen, Handelsnamen, Warenbezeichnungen u.s.w. in diesem Werk berechtigt auch ohne besondere Kennzeichnung nicht zu der Annahme, dass solche Namen im Sinne der Warenzeichen- und Markenschutzgesetzgebung als frei zu betrachten wären und daher von jedermann benutzt werden dürften.

Verlag: Südwestdeutscher Verlag für Hochschulschriften Aktiengesellschaft & Co. KG
Dudweiler Landstr. 99, 66123 Saarbrücken, Deutschland
Telefon +49 681 37 20 271-1, Telefax +49 681 37 20 271-0
Email: info@svh-verlag.de
Zugl.: Köln, Universität, Dissertation, 2009

Herstellung in Deutschland:
Schaltungsdienst Lange o.H.G., Berlin
Books on Demand GmbH, Norderstedt
Reha GmbH, Saarbrücken
Amazon Distribution GmbH, Leipzig
ISBN: 978-3-8381-1357-9

Imprint (only for USA, GB)
Bibliographic information published by the Deutsche Nationalbibliothek: The Deutsche Nationalbibliothek lists this publication in the Deutsche Nationalbibliografie; detailed bibliographic data are available in the Internet at http://dnb.d-nb.de.

Any brand names and product names mentioned in this book are subject to trademark, brand or patent protection and are trademarks or registered trademarks of their respective holders. The use of brand names, product names, common names, trade names, product descriptions etc. even without a particular marking in this works is in no way to be construed to mean that such names may be regarded as unrestricted in respect of trademark and brand protection legislation and could thus be used by anyone.

Publisher: Südwestdeutscher Verlag für Hochschulschriften Aktiengesellschaft & Co. KG
Dudweiler Landstr. 99, 66123 Saarbrücken, Germany
Phone +49 681 37 20 271-1, Fax +49 681 37 20 271-0
Email: info@svh-verlag.de

Printed in the U.S.A.
Printed in the U.K. by (see last page)
ISBN: 978-3-8381-1357-9

Copyright © 2010 by the author and Südwestdeutscher Verlag für Hochschulschriften Aktiengesellschaft & Co. KG and licensors
All rights reserved. Saarbrücken 2010

Abstract

In the unglaciated areas of Antarctica, lake sediments act as archives of the regional environmental and climatic history. In most cases, the records are restricted to the Holocene. Amongst the few exceptions are lakes in the McMurdo Dry Valleys, southern Victoria Land, which are known to have remained mostly ice-free during the Last Glacial Maximum.

Within the scope of an U.S.-American-German expedition in austral summer 2002/2003, several sediment cores were recovered from the three major lakes in the Taylor Valley: lakes Fryxell, Hoare, and Bonney. In order to reconstruct the late Quaternary regional environmental and climate history, sedimentological, biogeochemical, mineralogical, and chronological investigations were conducted on the sediment sequences recovered from Lake Hoare (core Lz1020) and East Lake Bonney (core Lz1023) within the scope of this thesis.

Sediment cores from Lake Hoare with a maximum length of 2.3 m mainly consist of coarse-grained material and penetrate back into the late Weichselian, when Taylor Valley was occupied by the large proglacial Lake Washburn. This lake was dammed by the advanced Ross Sea ice sheet at the valley outlet and was mainly fed by meltwater of the ice sheet. During the Pleistocene-Holocene transition, enhanced evaporation led to a significant lake level drop of Lake Washburn. The Lake Hoare record additionally shows that in course of this event, Lake Washburn desiccated to a very low level, with subaerial conditions at the coring site of Lz1020. After the final retreat of the Ross Sea ice sheet during the early Holocene, Taylor Valley was occupied by remnants of Lake Washburn. Environmental conditions comparable to those of today, with an advanced Canada Glacier separating lakes Hoare and Fryxell, established during the Mid-Holocene.

A 2.7 m long core recovered from East Lake Bonney mainly consists of a halite crust. Variations in the properties of the salt crystals and of clastic components embedded in the salts imply environmental changes over time. New paleoenvironmental insights provided by this record are the evidence for enhanced evaporation in the late Holocene, which led to the precipitation of the more than 2 m thick salt crust. This event was followed by a lake level rise, caused by inflowing meltwaters refilling the basin. As a result of the establishment of a freshwater lens at the top of East Lake Bonney, a perennial ice cover was formed in the recent past.

This study shows that the investigated lake sediment records provide crucial information about the late Quaternary environmental history of Taylor Valley, but should be interpreted in context with ice core records, terrestrial, and marine archives for a better understanding of the regional paleoenvironment, and paleoclimate.

Zusammenfassung

In den eisfreien Gebieten der Antarktis fungieren Seesedimente als Archive der regionalen Umwelt- und Klimageschichte. In den meisten Fällen ist ihre zeitliche Reichweite jedoch auf das Holozän beschränkt. Eine der wenigen Ausnahmen stellen die Seen in den Trockentälern der McMurdo-Region im Süd-Viktorialand dar, die während des Letzten Glazialen Maximums überwiegend eisfrei geblieben sind.

Im Rahmen einer U.S.-amerikanisch-deutschen Expedition im Südsommer 2002/2003 wurden Sedimentkerne aus den drei größten Seen des Taylor Valley, Fryxell, Hoare und Bonney, geborgen. Mit dem Ziel, die spätquartäre regionale Umwelt- und Klimageschichte zu rekonstruieren, sind die Sedimentsequenzen des Lake Hoare (Kern Lz1020) und des östlichen Beckens des Lake Bonney (Kern Lz1023) im Rahmen dieser Arbeit hinsichtlich ihrer sedimentologischen, biogeochemischen, mineralogischen und chronologischen Eigenschaften untersucht worden.

Die Sedimentkerne des Lake Hoare, die eine Maximallänge von 2,3 m erreichen und überwiegend aus grobkörnigen Material bestehen, reichen zurück bis in die späte Weichselzeit, als das Taylor Valley von dem großen proglazialen Lake Washburn erfüllt war. Dieser See wurde durch den vorgerückten Eisschild, der sich im heutigen Rossmeer befand, aufgestaut und vorwiegend durch dessen Schmelzwässer gespeist. Während des Übergangs vom Pleistozän zum Holozän führte verstärkte Evaporation zu einem signifikanten Seespiegelabfall des Lake Washburn. Der Lake Hoare Rekord zeigt außerdem, dass im Verlauf dieses Ereignisses die Bohrlokation des Kerns Lz1020 trocken fiel und sich subaerische Sedimentationsbedingungen einstellten. Nach dem endgültigen Rückzug des Rossmeer-Eisschilds während des frühen Holozäns verblieben nur Überreste des früheren Lake Washburn im Taylor Valley. Mit dem Vorrücken des Canada-Gletschers im mittleren Holozän wurde der Lake Hoare vom Lake Fryxell getrennt und es stellten sich Umweltbedingungen ein, die vergleichbar mit den heutigen sind.

Aus dem östlichen Becken des Lake Bonney konnte ein 2,7 m langer Sedimentkern geborgen werden, der überwiegend aus einer Halitkruste besteht. Variationen in den Eigenschaften sowohl der Salzkristalle als auch der klastischen Komponenten, die in die Salze eingebettet sind, deuten auf Umweltveränderungen in der Vergangenheit hin. Neue Erkenntnisse bezüglich der Paläoumwelt ergeben sich aus diesem Rekord anhand von Belegen für eine verstärkte Evaporationsphase im Spätholozän, die zu der Ablagerung der mehr als 2 m mächtigen Salzkruste führte. Auf diese Phase folgend stieg der Seespiegel des östlichen Lake Bonney an, was auf einen verstärkten Zufluss von Schmelzwässern zurückzuführen ist. Dadurch bildete sich eine Frischwasserlinse auf dem östlichen Lake Bonney aus, die dafür verantwortlich ist, dass sich in jüngster Vergangenheit eine permanente Eisdecke auf dem See entwickeln konnte.

Die vorliegende Arbeit zeigt, dass die untersuchten Seesedimentsequenzen wichtige Informationen über die spätquartäre Umweltgeschichte des Taylor Valley liefern, aber dass diese auch im Zusammenhang sowohl mit Eiskerndaten als auch terrestrischen und marinen Archiven interpretiert werden sollten, um die regionale Paläoumwelt und das regionale Paläoklima besser zu verstehen.

Table of contents

List of figures..III
List of tables IV
List of abbreviations..V

1 Introduction..1

2 Background – Current state of research 3
   2.1 Late Quaternary climate history of the McMurdo Dry Valleys...................................3
   2.2 Late Quaternary environmental history of the McMurdo Dry Valleys........................6
      2.2.1 Glacial history 6
      2.2.2 Proglacial lakes..9
      2.2.3 Holocene lake history 12

3 Study area...14
   3.1 Lakes of Taylor Valley...15
   3.2 Geology..16
   3.3 Climate..17

4 The Lake Hoare record 18
   4.1 Introduction...18
   4.2 Lake Hoare..19
   4.3 Material and methods 21
   4.4 Results and discussion 25
      4.4.1 Lithology..25
         4.4.1.1 Core description...26
         4.4.1.2 Core correlation and discussion...28
      4.4.2 Biogeochemistry..31
      4.4.3 Mineralogy 34
      4.4.4 Chronology..35
         4.4.4.1 Radiocarbon ages..35
         4.4.4.2 Ages of units...37
         4.4.4.3 Reservoir effect changes 38
   4.5 Implications for the environmental history of Taylor Valley 40
      4.5.1 Unit I (~17,000-14,000 years BP) – Lake Washburn 41

I

    4.5.2 Unit II (~14,000-11,000 years BP) – Lake Washburn evaporation 42

    4.5.3 Unit III (~11,000-9000 years BP) – Lake Washburn desiccation 43

    4.5.4 Unit IV (~9000 years BP to present) – Holocene lake history 46

  4.6 Conclusions 48

5 The East Lake Bonney record 50

  5.1 Introduction 50

  5.2 Lake Bonney 51

  5.3 Material and methods 53

  5.4 Results and discussion 55

    5.4.1 Stratigraphy 55

      5.4.1.1 Unit 1 56

      5.4.1.2 Unit 2 58

      5.4.1.3 Unit 3 58

      5.4.1.4 Unit 4 59

    5.4.2 Clastic fraction 59

    5.4.3 Salt fraction 61

    5.4.4 Chronological considerations 63

  5.5 Paleoenvironmental implications 65

  5.6 Conclusions 69

6 Lacustrine history of Taylor Valley - Synthesis 70

  6.1 Introduction 70

  6.2 Assessment of proxies for paleoenvironmental reconstructions 71

  6.3 Reconstruction of the late Quaternary environmental history of Taylor Valley, Antarctica 74

  6.4 Implications for the Holocene climate history of East Antarctica 81

  6.5 Conclusions 83

7 Summary and final remarks 85

References 87

List of figures

Figure 1. Climate history of southern Victoria Land, illustrated by $\delta^{18}O$ values, sodium (Na) and ^{10}Be concentrations from Taylor Dome ice record, and climatic implications from zoological studies. 4

Figure 2. Index map of western Ross Sea and of the southern Scott Coast 7

Figure 3. Flowlines and surface contours (elevation in meters) of the grounded ice sheet that deposited Ross Sea drift in the McMurdo Sound region 8

Figure 4. Environmental history of McMurdo Dry Valleys 10

Figure 5. Glacial Lake Washburn 12

Figure 7. McMurdo Dry Valleys, Antarctica 14

Figure 8. Landsat-7 satellite image of Taylor Valley, McMurdo Dry Valleys, Antarctica, indicating the location of the most important lakes and glaciers 15

Figure 9. Geology of Taylor Valley 16

Figure 10. Study site - Lake Hoare 19

Figure 11. (a) Lake Hoare bathymetry and (b) waterprofile 20

Figure 12. Flowchart of methods for analyzing core Lz1020 from Lake Hoare. 22

Figure 13. Lithology of Lake Hoare cores 25

Figure 14. SEM photographs of diatoms in the microbial mats of the surface sediment from Lake Hoare 26

Figure 15. Characterization of core Lz1020 from Lake Hoare 32

Figure 16. SEM photograph of aragonite needles in core Lz1020-4, 170-172 cm depth, from Lake Hoare 33

Figure 17. Age-depth distribution of core Lz1020 from Lake Hoare. 37

Figure 18. Reconstruction of the environmental history of eastern Taylor Valley, Antarctica 40

Figure 19. Correlation of core Lz1020 from Lake Hoare and core Lz1021 from Lake Fryxell 42

Figure 20. Lake Hoare bathymetry map showing current and retreated glacier positions, along with remnant ponds (with approximate depths) and streams 44

Figure 21. Study site - East Lake Bonney 51

Figure 22. (a) Lake Bonney bathymetry and (b) waterprofile of East Lake Bonney 52

Figure 23. Flowchart of methods for analyzing core Lz1023 from East Lake Bonney. 54

Figure 24. Diffractogram and SEM photo of salt crystal from core Lz1023, 150 cm depth. 56

Figure 25. Characterization of core Lz1023 from East Lake Bonney 57

Figure 26. Total organic carbon versus total sulfur values from East Lake Bonney core Lz1023 61

Figure 27. Radiocarbon ages of DIC, and DIC concentrations in East Lake Bonney (ELB) 64

Figure 28. Chronology of events at Lake Bonney during the Holocene 67

Figure 29. Taylor Valley. 70

Figure 30. The Lake Fryxell record 74

Figure 31. Reconstruction of the late Quaternary environmental history of eastern Taylor Valley....... 75

Figure 31. Radiocarbon ages of organic remains from ancient deltas versus their altitude in Taylor Valley.. 77

Figure 32. Overview of the late Quaternary environmental history of Taylor Valley. 80

Figure 33. Holocene climate history of East Antarctica inferred from different archives 82

List of tables

Table 1. Overview of assumed refilling ages for East Lake Bonney. 13

Table 2. Properties of major Taylor Valley lakes. 16

Table 3. Climate statistics for Taylor Valley 17

Table 4. Radiocarbon ages of the humic acid free fraction (HAF), and the humic acid fraction (HA) from bulk sediment samples of core Lz1020, Lake Hoare.. 24

Table 5. Radiocarbon dates of carbonates from a bulk sediment sample of core Lz1023, East Lake Bonney.. 55

Table 6. Radiocarbon data of DIC in East Lake Bonney. 64

Table 7. U/Th ages of former ELB cores 65

List of abbreviations

AAA	acid-alkali-acid treatment (extraction method for radiocarbon dating)
AAS	atomic absorption spectrometry
amph	amphiboles
AMS	atomic mass spectrometry
BP	before present
CTDS	corrected total dissolved solids
DIC	dissolved inorganic carbon
DOC	dissolved organic carbon
DVDP	Dry Valleys Drilling Project
EAIS	East Antarctic Ice Sheet
ELB	East Lake Bonney
fsp	feldspars
GSD	grain-size distribution
HA	humic acid
HAF	humic acid free
LGM	last glacial maximum
m a.s.l.	meters above sea level
MCM LTER	McMurdo Long-term Ecological Research Program
MIS	Marine Isotope Stage
POC	particulate organic carbon
psu	practical salinity unit
px	pyroxenes
qz	quartz
RIS	Ross Sea ice sheet
rre	recent reservoir effect
SEM	scanning electron microscope
std	standard
TC	total carbon
TDS	total dissolved solids
TIC	total inorganic carbon
TN	total nitrogen
TOC	total organic carbon
TS	total sulfur
WAIS	West Antarctic Ice Sheet
WLB	West Lake Bonney
XRD	X-ray diffractometry

1 Introduction

In times of intensive discussions about climate change and its future impacts, reconstruction of the past climate becomes more important as a basis for climate forecasting and modeling. Within paleoclimatic and paleoenvironmental studies, Antarctica plays an important role, since the human impact on its environment is negligible. In this context, ice core records, e.g. from Vostok (Petit et al., 1999), from Dome C (EPICA, 2004) or from Taylor Dome (Steig et al., 2000), provide important information about the past climate. Sedimentary records, both marine and terrestrial, can offer additional information about regional paleoenvironmental histories (e.g., Berkman et al., 1998; Ingólfsson, 2004).

Our present knowledge of the climatic history along the Antarctic coastline, and the climate impact on the glacial and preglacial environment is mainly based on investigations of sediments deposited on the vast continental shelves and in the restricted unglaciated coastal areas (Ingólfsson, 2004). In the latter areas, lake sediment successions turned out to be promising archives of climate due to their commonly continuous formation, rather high sedimentation rates, relatively good age control by radiocarbon dating, and high sensitivity to climatic and environmental changes (e.g., Verleyen et al., 2004; Wagner et al., 2006). The onset of lacustrine sedimentation in most areas is restricted to the Holocene, commencing only after the ice retreat at the end of the late Weichselian (Marine Isotopic Stage 2) glaciation (Hodgson et al., 2004). In contrast, investigations on the Larsemann Hills provide evidence that parts of the Brokness peninsula remained ice-free during the last glaciation and some lakes have existed since 44,000 years BP (Hodgson et al., 2001). The McMurdo Dry Valleys in the vicinity of the McMurdo Sound, Ross Sea, remained mostly unglaciated during the last glaciation (Doran et al., 1994). Hence, lake sediment sequences from this region can provide a record reaching back into the last glacial period.

In austral summer 2002/2003, an expedition to the McMurdo Dry Valleys, southern Victoria Land, Antarctica, took place in order to recover several meters long lake sediment cores from the three major lakes of Taylor Valley for the first time. The aim of the collaboration project between the workgroup of Martin Melles from the University of Leipzig at that time, now at the University of Cologne, and the working group around Peter Doran and Fabien Kenig from the University of Illinois at Chicago, U.S.A. is the reconstruction of the late Quaternary environmental history of Taylor Valley by a multidisciplinary study of lake sediments. The current thesis is part of the project, which was funded by the German Research Foundation (Deutsche Forschungsgemeinschaft, DFG; Project no. ME 1169/11) within the priority program 1158 "Antarktisforschung".

The study of the recovered lake sediment sequences is focused on the following points:

- creating an independent chronology,
- reconstruction of the late Quaternary environmental and climate history of Taylor Valley,
- comparing the results with those from Antarctic ice core, marine, and terrestrial records of the region and other coastal ice-free regions of Antarctica,
- figuring out of regional peculiarities of environmental changes and indications for their causes,
- contribution to the understanding of circum-Antarctic similarities and differences,
- better understanding of ocean-land-ice interactions.

Preliminary studies on the sediment sequence recovered from Lake Fryxell (Wagner et al., 2006) already demonstrated that the cores could be considered as unique material for reconstructing the environmental history in a coastal region of Antarctica. This thesis focuses on chronological, sedimentological, mineralogical, and biogeochemical studies of the sediment sequences recovered from Lake Hoare and East Lake Bonney.

Providing a better understanding of the region and its history, chapter 2 summarizes the present knowledge about the climate and environmental history of the McMurdo Dry Valleys, and chapter 3 gives an overview of the study area. The investigations on the Lake Hoare record and their results are presented in chapter 4, giving a detailed overview of the late Quaternary environmental history of eastern Taylor Valley. The sediment sequence recovered from East Lake Bonney is characterized and discussed in chapter 5. A concluding consideration and assessment of the Taylor Valley lake sediment sequences as archives for reconstructions of climate and environmental history is subject of chapter 6, followed by a final summary in chapter 7.

2 Background – Current state of research

Antarctica is the southernmost continent of the Earth, lying almost entirely south of the Antarctic Circle. The continent is geologically and glaciologically divided by the Transantarctic Mountains into the Eastern and Western Antarctic. Only 0.4 % of the continent is presently ice free (SCAR, 2007). Most of these ice-free areas, called oases, are located along the coast of East Antarctica. The McMurdo Dry Valleys region is the largest ice-free area in Antarctica and is said to have been largely ice-free during the last glacial period (Hendy, 2000a). According to this, lake sediment sequences obtained from this region can provide records reaching back into the last glacial. This chapter gives an overview of the late Quaternary environmental and climate history of the McMurdo Dry Valleys inferred from different records investigated in earlier studies, to outline the complexity of the environment of Taylor Valley and its history as a background for the interpretation of the lake sediment sequences investigated within the scope of this thesis.

2.1 Late Quaternary climate history of the McMurdo Dry Valleys

Due to its proximity to the dry valleys and the Ross Sea, the ice core record recovered from Taylor Dome provides the best information about the climate history of the region. Taylor Dome (77°47'47''S, 158°43'26''E; 2365 m a.s.l.) is a local ice cap at the margin of the East Antarctic ice sheet (EAIS), located ca. 150 km inland from the Ross Sea Embayment. An ice-core of 554 m in length covers the entire Weichselian and the Holocene (Steig et al., 2000). Past climate conditions are deduced from several proxies (Figure 1). $\delta^{18}O$ is an indicator for temperature. The ^{10}Be method can be used for estimation of past accumulation rates, whereby it has to be considered: the higher the ^{10}Be values are, the lower the accumulation rate is. Additionally, aerosol loading can also give indications for atmospheric circulation patterns in the region. The Taylor Dome record with its greater variance of temperature, snow accumulation and aerosol concentration compared to more continental ice core records reflects the significant variability in atmospheric circulation and air mass moisture content in coastal areas. These factors are strongly associated with the dynamics of the Ross Sea ice sheet. Additionally, the Taylor Dome record is in its character similar to Northern hemisphere climate changes, which can be traced back to the sensitivity of the western Ross Embayment to North Atlantic climate variability (Steig et al., 1998; Steig et al., 2000; Grootes et al., 2001).

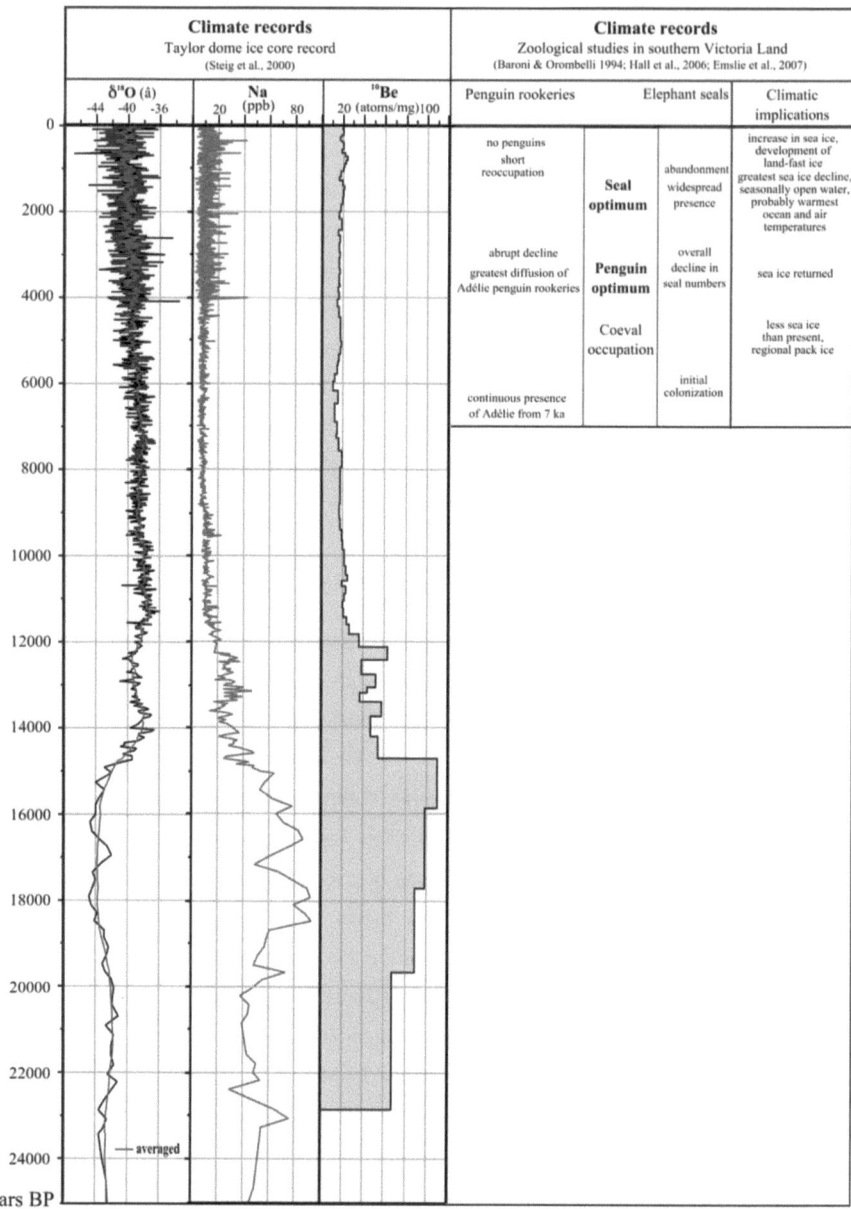

Figure 1. Climate history of southern Victoria Land, illustrated by $\delta^{18}O$ values, sodium (Na) and ^{10}Be concentrations from Taylor Dome ice record, and climatic implications from zoological studies.

During the last glacial period, $\delta^{18}O$ values are generally low indicating cold temperatures (Steig et al., 2000). The expansion of Ross Sea Ice Sheet had an effect on the dynamics of the low-pressure system in the Ross Sea, which shifted northward and was therefore responsible for drier climatic conditions with very low accumulation rates (high ^{10}Be values) (Steig et al., 2000). Increased aerosol loading during glacial times is the result of an extreme aridity and windiness caused by the strengthened atmospheric circulation systems more influenced by the Polar Plateau and with outflowing catabatic winds and a strong zonal circulation (Grootes et al., 2001). The coldest temperatures occur during the last glacial maximum between 20,000 and 16,000 years BP, which is interrupted by a smaller warming event at ~17,000 years BP (Figure 1). The Taylor Dome record shows a sharp decrease in the ^{10}Be values and a strong increase in $\delta^{18}O$ values at ca. 15,000 years BP, indicating a rapid transition from glacial to interglacial conditions, which peaks at ~14,000 years BP (Figure 1). This period is followed by a prominent cold period between ~13,000 and 12,000 years BP, which Steig et al. (1998) synchronize with the Younger Dryas cooling known from the North Atlantic region. From ca. 12,000 years BP, $\delta^{18}O$ values of the Taylor Dome ice record are relatively high, indicating warmer than present temperatures, and accumulation rates are increased (Figure 1). Following the mid-Holocene temperature rise, a general cooling trend is significant in particular since 6000 years BP (Steig et al., 2000). The high accumulation rates and increasing Na concentrations (Figure 1) throughout the middle and late Holocene can be explained by the same dominant circulation patterns as prevailing today, which are influenced by open water conditions in the Ross Sea after the final retreat of the Ross Sea Ice Sheet (Steig et al., 2000). The atmospheric circulation patterns with dominating cyclonic systems support a high moisture supply to Taylor Dome by bringing precipitation from the Ross Sea to Taylor Dome (Grootes et al., 2001).

Information about climate change during the Holocene climate can also be derived from zoological studies (Figure 1) by dating of abandoned penguin rookeries and elephant seal colonies along the coast of southern Victoria Land, since penguins and elephant seals are sensitive to special climatic and environmental conditions (Baroni and Orombelli, 1994a; Emslie et al., 2007; Hall et al., 2006b). The "Penguin Optimum" between 4000 and 3000 ^{14}C years BP (Figure 1) can be related to climate conditions in the Ross Sea, which were characterized by more sea ice. Investigations of Hall et al. (2006) on the Holocene elephant seal distribution assumed a climate optimum between 2300 and 1100 ^{14}C years BP ("Seal Optimum", Figure 1), since elephant seal colonies expanded significantly during this period due to the greatest sea-ice decline during Holocene times.

2.2 Late Quaternary environmental history of the McMurdo Dry Valleys

The McMurdo Dry Valleys are the largest ice-free area in Antarctica today. The landscape was essentially formed before the middle Miocene, when fluvial planation and downcutting created the main features of the present-known valley system. Following these denudation and erosion processes, the landscape experienced a substantial subsidence of estimated 400 m, whereby the valleys were flooded by the sea (Denton et al., 1993). In cores of the Dry Valley Drilling Project (DVDP), sedimentation in a 600-900 m deep fjord was reconstructed for the Pliocene (Elston and Bressler, 1981). Tectonic processes lead to an uplift of approximately 300 m, and the valleys' floors were successively set to present-day elevation. Glacial superimposing of the region is documented since at least the Pliocene. Several drift deposits document incursions by continental glacier ice and by shelf ice (Kyle, 1981).

2.2.1 Glacial history

In the McMurdo Dry Valleys, three major glacier systems are important for the glacial formation of the landscape: the Ross Sea ice shelf, which is influenced by the West Antarctic Ice Sheet (WAIS), the East Antarctic Ice Sheet (EAIS) with its draining glacier systems and local alpine glaciers of the Transantarctic Mountains (Denton et al., 1989). Drift deposits in Taylor Valley derive from all three systems.

The eastern part of Taylor Valley contains thick glacial deposits (called Ross Sea drift I) mainly originating from an advance of the Ross Sea ice sheet (RIS) in McMurdo Sound during the last glacial period. Thickening and expansion of the WAIS led to an advance of the RIS and its grounding in McMurdo Sound and along the Scott Coast (Conway et al., 1999) (Figure 2).

Investigations on marine and terrestrial archives show different reconstructions of the maximum extent of the RIS for the last LGM. After Stuiver et al. (1981), the grounding line of the ice sheet reached to or close to the edge of the continental shelf (between 72 and 73°S, Figure 2). Licht et al. (1996; 1999) found evidence in marine sediments that the groundling line advanced to ca. 100 km south of Coulman Island (~74°S, Figure 2), whereas Domack et al. (1999) and Shipp et al. (1999) assume the maximum extent of the Ross Sea Ice Sheet in the western Ross Sea north of Coulman Island (Figure 2). This is supported by investigations on abandoned penguin rookeries (Baroni and Orombelli, 1994a; Emslie et al., 2007). It is assumed that the ice edge was located near Coulman Island during the LGM at least until 13,000 ^{14}C years BP, since indications for breeding south of 74°S are lacking (Emslie et al., 2007). However, terrestrial archives (Hall and Denton, 2000b) and modeling (Denton and

Hughes, 2000) indicate that the ice lobe reached thicknesses of more than 500 m in McMurdo Sound. Evidence of kenyte erratics and volcanic material originating from Ross Island in the continental valleys argue for a flowline of the ice sheet as shown in Figure 3. The significance of an extended Ross Sea ice sheet (RIS) is its influence on the regional climate and the environment of the dry valleys. Due to the occupation of a large sector of the Ross Sea by a thick ice lobe, the climate was drier, what can be deduced from low accumulation rates at Taylor Dome (Steig et al., 2000). Additionally, the RIS blocked the mouths of the valleys, whereby large proglacial lakes (chapter 2.2.2) could be dammed and filled the valleys, fed by meltwater of the ice sheet, up to 300 meters.

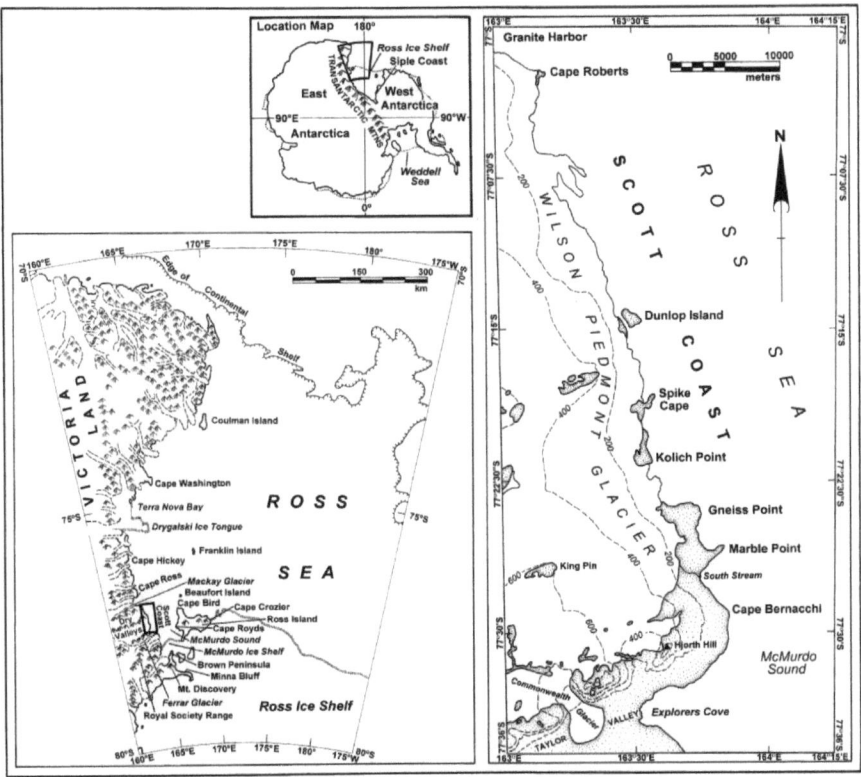

Figure 2. Index map of western Ross Sea and of the southern Scott Coast. Contours are in meters a.s.l. (Hall and Denton, 2000a).

The Ross Sea ice sheet (RIS) and ice shelf were retreating after the LGM at an average rate of 100 m/year (Bindschadler, 1998; Domack et al., 1999), whereas the ice shelf edge receded with a temporal delay (Shipp et al., 1999). The retreat of the RIS most likely started at the glacial-interglacial transition about 12,000 years ago. Reconstructions of the deglaciation history based on marine sediment records show slight temporally differences in its succession. After Licht et al. (1996) and Domack et al. (1999), the grounding line retreat passed the Drygalski Trough by ca. 11,000 ^{14}C years BP (Figure 2). On the other hand, McKay et al. (2008) found evidence for grounding line retreat to Ross Island between 11,000 and 10,000 ^{14}C years BP. Terrestrial archives from the dry valleys indicate the retreat of the RIS from McMurdo Sound between ca. 9400 and 7600 ^{14}C years BP (Conway et al., 1999; Hall and Denton, 2000b). Raised beaches along the Scott Coast (Figure 2) imply its deglaciation about 6600 ^{14}C years BP with a relative sea level change of ca. 32 m since that time (Hall and Denton, 1999). However, in the western Ross embayment, the edge of the ice shelf was pinned to the south of Ross Island, probably since ~9000 ^{14}C years BP (McKay et al., 2008).

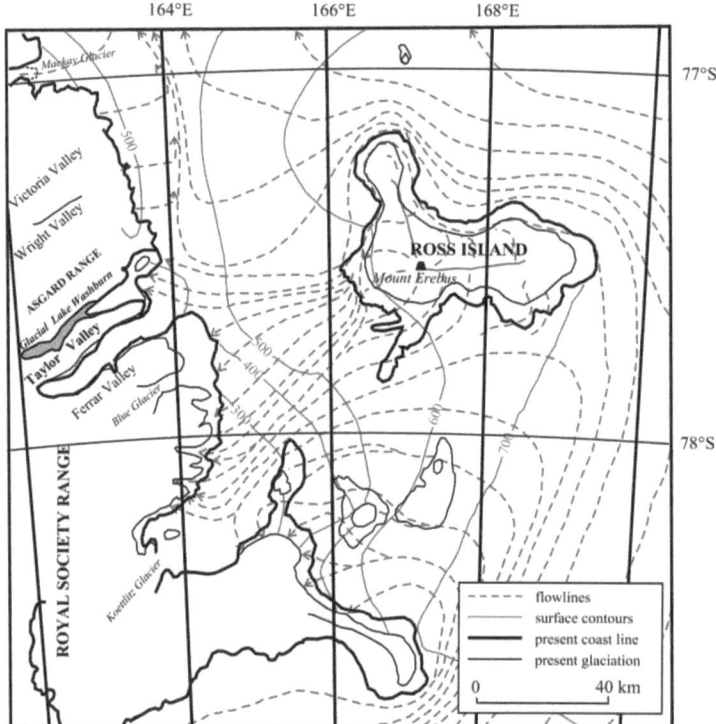

Figure 3. Flowlines and surface contours (elevation in meters) of the grounded ice sheet that deposited Ross Sea drift in the McMurdo Sound region (after Stuiver et al., 1981; Denton and Hughes, 2000; Denton and Marchant, 2000).

Today, alpine glaciers can be found at the slopes in the dry valleys, partly flowing into the deeper parts of the valley. The upper part of Taylor Valley is occupied by Taylor Glacier, which is draining the Taylor Dome (Hendy et al., 1979). Moraines and other drift deposits indicate several glacier advances, particularly during interglaciations (Hendy et al., 1979; Denton et al., 1993). When the Ross Sea was free of ice sheets, the climate was more humid with higher precipitation in the adjacent dry valleys. Thus, an elevated accumulation in the Taylor Dome region as well as in the area of the alpine glaciers led to a positive mass balance (Higgins et al., 2000b). According to this, Taylor Glacier and the alpine glaciers have been advancing during the Holocene (Alpine I drift) and are at their maximum positions today (Figure 4). During the last glacial, these glacial system were in a retreated position due to hyperarid climate conditions (Denton et al., 1989).

In summary, Taylor Valley contains glacial drift deposits originating from three glacial systems. Expansion of Taylor Glacier as an outlet glacier of the EAIS and local alpine glaciers occurred specifically during global interglaciations, and glacial recession during global glaciations (Higgins et al., 2000b). During the last glacial, the RIS advanced in the western Ross Embayment and was responsible for a damming of large proglacial lakes by plugging the valleys' mouths with ice (Stuiver et al., 1981) (Figure 4).

2.2.2 Proglacial lakes

Well-preserved geomorphological evidence for the existence of proglacial lakes exists for the last glaciation. Radiocarbon dated (glacio-)lacustrine sediments, paleodeltas and fossil strandlines give an impression of the lakes' extent and history, e.g. by reconstructing lake level fluctuations (Hall and Denton, 2000b; Hendy, 2000a; Hall et al., 2001; Hall et al., 2002). Damming of the proglacial lakes was caused by the grounded RIS in McMurdo Sound, which either terminated on land, like in Taylor Valley, or merged with the large outlet glacier systems, like the Wilson Piedmont Glacier, resulting in a thickening of these glaciers (Conway et al., 1999). Some of these lakes established an own characteristic mechanism of drift deposition, which can be described as a lake-ice conveyor belt (Hendy et al., 2000; Hall et al., 2006).

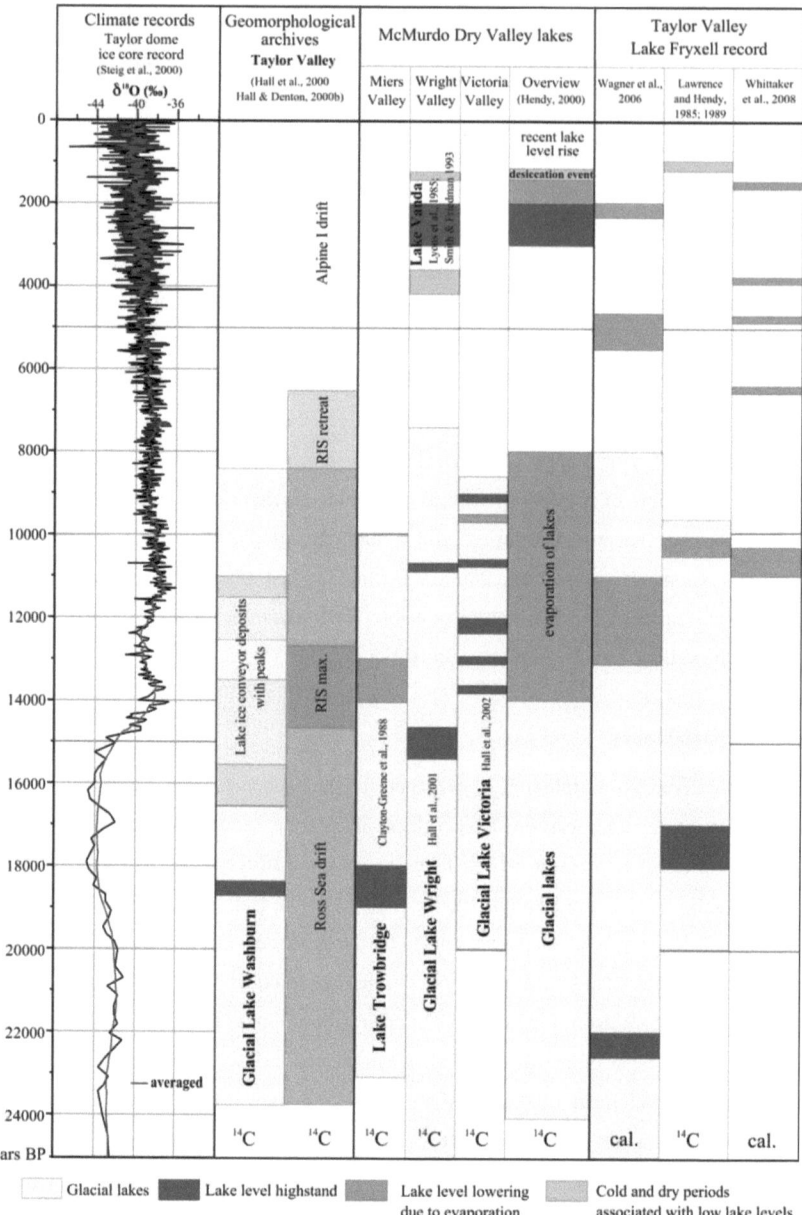

Figure 4. Environmental history of McMurdo Dry Valleys. For comparison, the climate history is illustrated by δ ^{18}O curve deduced from the Taylor Dome ice core record. Information about the environmental history is summarized from the most important studies mentioned in the text and includes geomorphological archives and lake sediment records of the McMurdo Dry Valleys.

Former occupancy of the dry valleys by large proglacial lakes is indicated by old lacustrine sediments. Isotopic signatures of lacustrine carbonates, which are assigned to these proglacial lakes, suggest that they were mainly fed by meltwater that derived from the ice sheet and large glacier systems in the dry valleys region, respectively (Hendy, 2000a).

Evidence for the existence of large proglacial lakes were found in Taylor Valley (Stuiver et al., 1981; Hall et al., 2000), Victoria Valley (Hall et al., 2002), Wright Valley (Hall et al., 2001), and Miers Valley (Clayton-Greene et al., 1988) in the dry valleys region (Figure 4). Similarities arise with regard to existence of the large proglacial lakes during last glacial maximum mostly until ca. 8000 ^{14}C years BP. Highstands occurred at several stages, mainly between 20,000 and 12,000 ^{14}C years BP (Figure 4). During these periods, meltwater supply was most likely increased despite a very cold and hyperarid climate. However, many cloudless and snowless days during the summer seasons were responsible for an elevated radiation induced glacial melting (Hall et al., 2001; 2002). Changing climate conditions can be deduced for the glacial-interglacial boundary documented in form of carbonate layers within the sediments of all these lakes (Figure 4). The former large lakes evaporated to smaller ones probably already before the final retreat of the RIS (Hendy, 2000a). The dates of last evidence for their existence coincide with the final disappearance of the RIS from the western Ross Sea Embayment (Figure 4).

In Taylor Valley, Glacial Lake Washburn was dammed by an advanced Ross Sea Ice Sheet (RIS), blocking the mouth of the valley (Figure 5). Moraines found near Hjort Hill indicate that the ice lobe reached an elevation of approximately 350 m a.s.l. at its maximum extent between 14,600 and 12,700 ^{14}C years BP (Hall and Denton, 2000b). Dated paleodeltas and fossil strandlines indicate that the lake existed at least between 23,800 and 8340 ^{14}C years BP (Figure 4) and underwent significant lake level oscillations (Hall and Denton, 2000b). Its maximum lake level was reached at ca. 18,500 ^{14}C years BP, proposed by a perched delta found at 314 m a.s.l. (Hall and Denton, 2000b). In addition, these authors detected two periods of increased lake ice conveyor deposition (Figure 4), which they ascribe to a very active conveyor, operating in the proglacial lake, or to a rapid lake level drop. The climate change after the LGM finally led to an evaporational lowering of Lake Washburn, as indicated by the occurrence of carbonate horizons, e.g. found in sediment sequences of Lake Fryxell (Lawrence and Hendy, 1989), and by the low altitudes of deltas originating from this period (Hall and Denton, 2000b) (Figure 31). Since there is no evidence for an outflow event related to the RIS retreat, it can be concluded that Glacial Lake Washburn was already evaporated to such a low level that only smaller isolated lakes existed in the separate drainage basins of Taylor Valley.

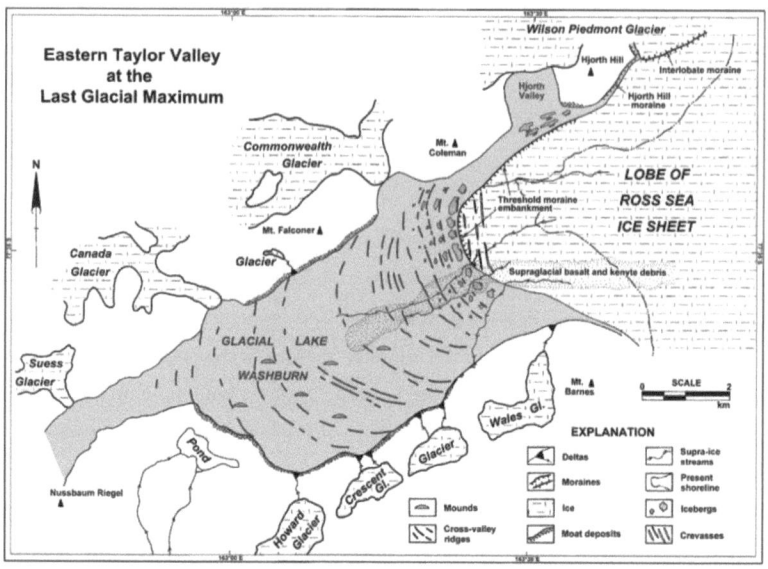

Figure 5. Glacial Lake Washburn (modified after Hall and Denton, 2000b).

2.2.3 Holocene lake history

During the Holocene, after the final retreat of the RIS from the mouth of Taylor Valley between 8340 and 6500 ^{14}C years BP (Hall and Denton, 2000b), the climate was getting warmer and more humid (Steig et al., 2000). As a result, alpine glaciers advanced in the course of the Alpine I drift. Their meltwaters formed the major source for the lakes, which constitute the remnants of the large glacial lakes (Hendy, 2000a).

During the Holocene, lake-level oscillations occurred, probably controlled by climate variations. However, geomorphological evidence of Holocene lake-level fluctuations is rare. Thus, reconstructions of the lakes' Holocene history are mainly based on short lake sediment cores and on diffusion models of chemical elements and isotopes obtained from recent lake water. Dated carbonate layers in sediment cores recovered from Lake Fryxell indicate several low lake levels during the Holocene (Wagner et al., 2006; Whittaker et al., 2008) (Figure 4). Based on radiocarbon dated paleodeltas, Hall (2003) suggest that lake-level highstands in the dry valleys occured at ~6000 and ~3000 cal. years BP. Fossil strandlines in the vicinity of Lake Vanda in Wright Valley (Smith and Friedman, 1993) indicate that the lake possibly reached a lake-level highstand between 3000 and 2000 ^{14}C years BP (Figure 4). Following this highstand, Lyons et al. (1998) suggest that the dry valley lakes experienced an evaporation event, based on diffusion models of elements and isotopic studies of the water columns of the present lakes. During this event, Lake Hoare is supposed to have been completely desiccated around 1200 years ago, whereas lakes Fryxell and Bonney are regarded as to have been evaporated to hypersaline ponds (Lyons et al., 1998) (Figure 4). Since then, lake levels were probably rising This is supported by recent observations since the beginning of the 20th century (Chinn, 1993).

Due to the extensive geochemical studies conducted on Lake Bonney waters, information about its Holocene history was obtained from diffusion models (Lyons et al., 1998; Poreda et al., 2004). Since both lobes of Lake Bonney are supposed to have rising lake levels, molecular diffusion can be observed in the chemical composition of the water bodies. It is assumed that Lake Bonney was evaporated to a low level in the earlier Holocene. The scenarios describe an initial flooding of the western lobe before 6000 years BP (Hendy et al., 1977). Around 3000 years BP, the lake level of the west lobe reached a highstand. This was likely supported by an advance of Taylor Glacier. As a result, lake waters overflowed the Bonney Riegel and spilled over into the east lobe (Poreda et al., 2004). The onset of the initial flooding of East Lake Bonney varies between the authors due to differences in the investigated elements and isotopes in the present water column (Table 1).

Table 1. Overview of assumed refilling ages for East Lake Bonney.

authors	parameter	assumed refilling age for East Lake Bonney
Poreda et al., 2004	He, Ne, Ar, N$_2$	3000 years
Lyons et al., 1998	δ^{18}O, δD, Cl$^-$, δ^{37}C	1000-1200 years
Matsubaya et al., 1979	δ^{18}O, δD, Cl$^-$	1200-2600 years
Hendy et al., 1977	δ^{18}O, NaCl	1800-4600 years and 400-750 years

3 Study area

Taylor Valley is located in the McMurdo Dry Valleys (Figure 6), the largest ice-free area in Antarctica in the southern part of Victoria Land with a combined area of approximately 4,800 km². The establishment of the dry valleys as a so-called Antarctic oasis can be traced back to the blocking of the ice-masses flow from the Polar Plateau by the Transantarctic Mountains, and the extreme aridity of the region (Fountain et al., 1999).

The region consists of several valleys, from which Taylor, Wright, and Victoria Valleys are the largest ones (Figure 6). The McMurdo Sound separates the continental mainland from the volcanic Ross Island. The Ross Sea is generally ice-free in this area, with sea ice forming only during the winter season. The McMurdo Ice shelf is the extensional part of the Ross Sea ice sheet, which is fed by the West Antarctic Ice Sheet (WAIS), and is presently pinned south of Ross Island (Figure 6).

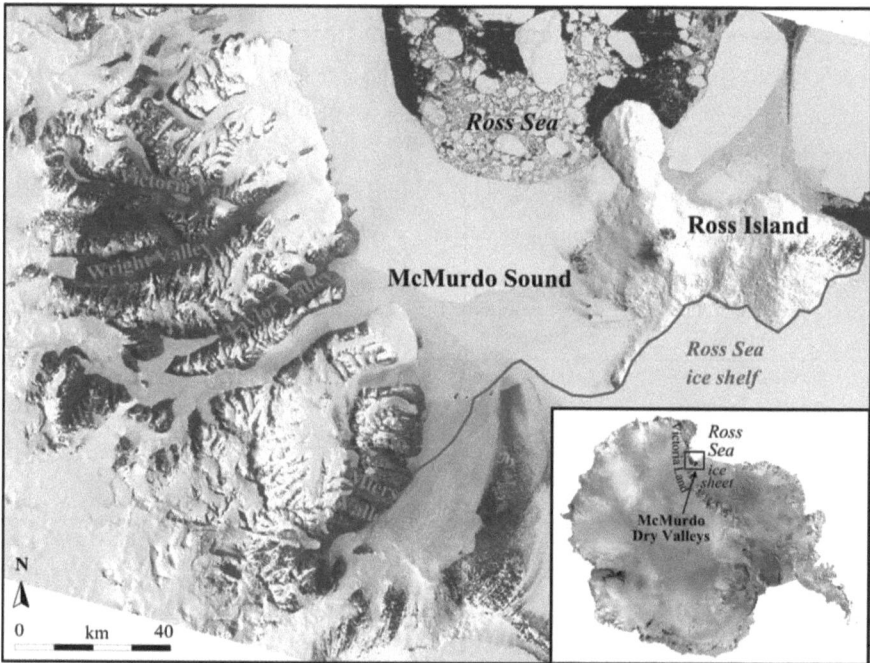

Figure 6. McMurdo Dry Valleys, Antarctica (U.S. Geological Survey, 2006; 2007).

The nearly east-west orientated Taylor Valley (Figure 7) is one of the largest valleys in the south of Dry Valleys region with ~35 km in length. Taylor Glacier occupies the upper valley in the west, in the east the valley borders to the Ross Sea. Mountains of the Asgaard Range and the Kukri Hills with heights of about 2000 m a.s.l. flank the valley to the north and to the south, respectively. The valley slopes are partly covered with alpine glaciers, e.g. Canada Glacier (Fountain et al., 1999) (Figure 7).

Figure 7. Landsat-7 satellite image of Taylor Valley, McMurdo Dry Valleys, Antarctica, indicating the location of the most important lakes and glaciers (U.S. Geological Survey, 2007). Blue points mark the location of the meteorological stations (data in Table 3). Coring locations are indicated by stars, red stars mark the cores investigated within the scope of this thesis.

3.1 Lakes of Taylor Valley

Taylor Valley consists of three basins (Figure 7): Explorers Cove, Fryxell, and Bonney basins. The Explorers Cove basin at the mouth of Taylor Valley is separated from the Fryxell basin by a sill with an elevation of ca. 78 m a.s.l.. A 116 m high threshold near the Suess Glacier separates Fryxell and Bonney basins in the western part of the valley. The Explorers Cove basin is currently occupied by the Ross Sea, while the deepest parts of the Fryxell and Bonney basins contain lakes. Lake Bonney with its eastern and western lobe is located in the upper, western drainage basin of the valley. Lakes Fryxell and Hoare are situated in the lower, eastern drainage basin and are separated by the Canada Glacier (Figure 7).

Depending on their location, the extent and the maximum water depth of these three lakes are varying (Table 2). The lakes all have a perennial ice cover of a few meters in thickness, but also contain unfrozen water. During the summer seasons, the ice covers occasionally melt along the shore and a narrow zone with open water forms the so-called moat (Doran et al., 1994).

Table 2. Properties of major Taylor Valley lakes (Lyons and Priscu, n.d.).

	Lake Hoare	East Lake Bonney	West Lake Bonney	Lake Fryxell
latitude	77° 38' S	77° 43' S	77° 43' S	77° 37' S
longitude	162° 55' E	162° 26' E	162° 17' E	163° 09' E
distance to sea (km)	15	25	28	9
elevation (m)	73	57	57	18
maximum length (km)	4.2	4.8	2.6	5.8
maximum width (km)	1	0.9	0.9	2.1
maximum depth (m)	34	37	40	20
surface area (km^2)	1.94	3.32	0.99	7.08
average ice thickness (m)	3.1 - 5.5	3.0 - 4.5	2.8 - 4.5	3.3 - 4.5
volume (x 10^6 m^3)	17.5	54.7	10.1	25.2

3.2 Geology

The crystalline basement in Taylor Valley consists of Precambrian to early Palaeozoic bedrock of the Granite Harbour Intrusive Complex and the Skelton Group (Figure 8), which are characterized by granitoids, gabbros, lamprophyres and metasediments, respectively (Angino et al., 1962; Cox and Allibone, 1991; Smillie, 1992). Sediments of the Paleozoic to early Mesozoic Beacon Super Group and dolerites of the Jurassic Ferrar Group occur in the western valley. Sporadic deposits of the Cenozoic McMurdo Volcanic Group are confined to the upper valley (Porter and Beget, 1981; Powell, 1981). The alkali volcanic complexes of this group are more dominant in the southern and eastern parts of McMurdo Sound (e.g., Mount Erebus on Ross Island). Quaternary sediments, overlying the bedrock, dominate the lower parts of Taylor Valley (Stuiver et al., 1981) (Figure 8).

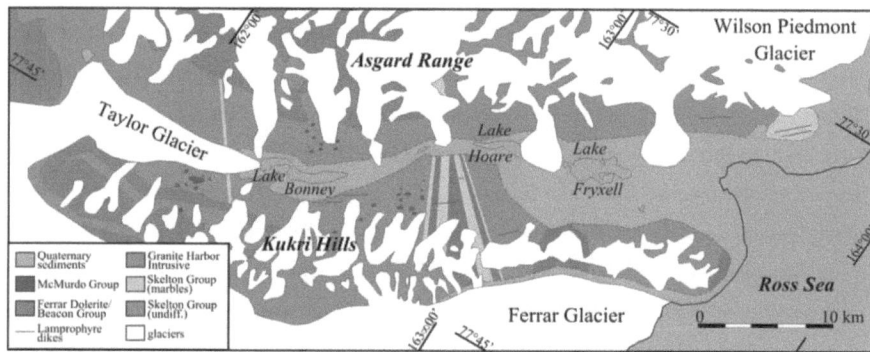

Figure 8. Geology of Taylor Valley (after Porter and Beget, 1981).

3.3 Climate

The climate of the McMurdo Dry Valleys is in general very cold and dry. However, spatial climatic variations can be observed in Taylor Valley (Table 3). The recorded mean annual valley bottom temperature in Taylor Valley ranges from -14.8 to -23.1°C (Doran et al., 2002a) (Table 3). The mean annual precipitation rate of <100 mm/year water equivalent mainly occurs in form of snow (Bromley, 1985). The low surface albedo, due to the darker colored bedrock and mostly lacking snow cover, together with dry katabatic winds descending from the Polar Plateau (Doran et al., 2002a), are responsible for the extremely dry conditions with mean ablation rates of 150-500 mm/year (Bromley, 1985; Clow et al., 1988). Easterly on-shore coastal winds are dominant in the austral summer. The highest wind speeds occur between March and November and are associated with westerly katabatic winds (Doran et al., 2002a; Doran et al., 2002b). The climate of the winter season is controlled by these wind regimes, whereas the summer climate is mainly influenced by the presence of solar radiation (Clow et al., 1988).

Table 3. Climate statistics for Taylor Valley (Doran et al., 2002a).

	Explorers Cove	Lake Fryxell	Lake Hoare	Lake Bonney
Period of station record				
start	21.11.1997	28.10.1987	12.12.1985	24.11.1993
end	25.01.2000	25.01.2000	24.01.2000	25.01.2000
Station elevation (m a.s.l.)	26	20	72	60
Distance from coast (km)	4	9	15	25
Air temperature (°C)				
avg. mean annual	-19.6	-20.2	-17.7	-17.9
max. mean annual	-19.2	-16.7	-14.8	-16.2
min. mean annual	-20.2	-23.1	-19.8	-19.1
absolute maximum	7.3	9.2	10.0	9.0
absolute minimum	-49.0	-60.2	-45.4	-47.9
Degree days above freezing				
mean	16.9	25.5	24.6	34.3
Relative humidity (%)				
avg. mean annual	74	69	66	62
absolute maximum	100	100	100	100
absolute minimum	<12.4	<15.5	<19.2	<12
Wind speed (m/s)				
avg. mean annual	2.5	3.1	2.8	3.9
absolute maximum	32.2	37.8	36.3	35.6

4 The Lake Hoare record

4.1 Introduction

Former reconstructions of the Late Quaternary environmental and climatic history of the McMurdo Dry Valleys are predominantly based on geomorphological features (e.g., Hall and Denton, 2000b), marine sediments (e.g., Cunningham et al., 1999), and deposits of penguin rookeries (e.g., Baroni and Orombelli, 1994b). The results of geomorphological studies indicate that large proglacial lakes have existed in the dry valleys during the last glacial period (Clayton-Greene et al., 1988; Hall et al., 2001; Hall et al., 2002). Investigations from Taylor Valley suggest the presence of a large, 300 m deep proglacial lake during the late Weichselian (Hall et al., 2000). This so-called "Lake Washburn" likely developed due to damming of Taylor Valley by an advanced Ross Sea Ice Sheet (RIS). The recent lakes of Taylor Valley (Figure 9) are believed to be remnants of Lake Washburn (Hendy, 2000a).

Previous investigations on Taylor Valley lakes focused on modern processes, with extensive biogeochemical and physical studies conducted within the scope of the McMurdo Long-Term Ecological Research (LTER) Program (e.g., Priscu, 1998). The history of the lakes, in contrast, remained little known. The use of lake sediments for the reconstruction of the environmental history has been limited to a few studies. Sediment cores were recovered from Lake Fryxell and contained evidence for evaporation events in the past (cf. Lawrence and Hendy 1985, 1989). Squyres et al. (1991) studied recent sedimentation processes in Lake Hoare. Investigations by Spaulding et al. (1997), and Doran et al. (1999) on short lake sediment cores from Lake Hoare provide information only about the most recent millennia.

Several meter long cores were recovered during an U.S.-German expedition in 2002 under the auspices of the McMurdo LTER (Wagner, 2003). The results obtained from a ca. 9 m long core (Lz1021), recovered from Lake Fryxell, for the first time provided largely continuous information about the lake history during the past ca. 48,000 years, with the development of modern Lake Fryxell from the ancient Lake Washburn (Wagner et al., 2006). The sediment sequences recovered from Lake Hoare, which is separated from Lake Fryxell by Canada Glacier (Figure 9), provide additional information on the late Quaternary environmental history of Taylor Valley. In this chapter, chronological, physical, and biogeochemical data are presented from the longest sediment sequence (core Lz1020) recovered from Lake Hoare to discuss the lacustrine history of the Fryxell basin for the late Weichselian and Holocene.

4.2 Lake Hoare

Lake Hoare (77° 38' S, 162° 53' E) with a surface area of 1.8 km^2 is located at an altitude of 73 m a.s.l. in a narrow section of the valley at the western end of the Fryxell basin (Figure 9; Table 2). Its bathymetry shows several sub-basins with an average water depth of 14.2 m (Schmok and Waddington, 1996) (Figure 10a). The deepest part (~34 m) and also the maximum width (~1 km) are located in the north-eastern part of the lake. Here, the tongue of alpine Canada Glacier forms a natural barrier (Wharton et al., 1989).

The thickness of the perennial ice cover of Lake Hoare is 3.1 to 5.5 m on average (Table 2). The ice cover is characterized by a rough surface. High amounts of aeolian sediment are trapped on the ice cover, most likely because down-valley blowing winds are blocked by Canada Glacier and thus lose their transport energy in front of the glacier (Wharton et al., 1989; Doran et al., 1994).

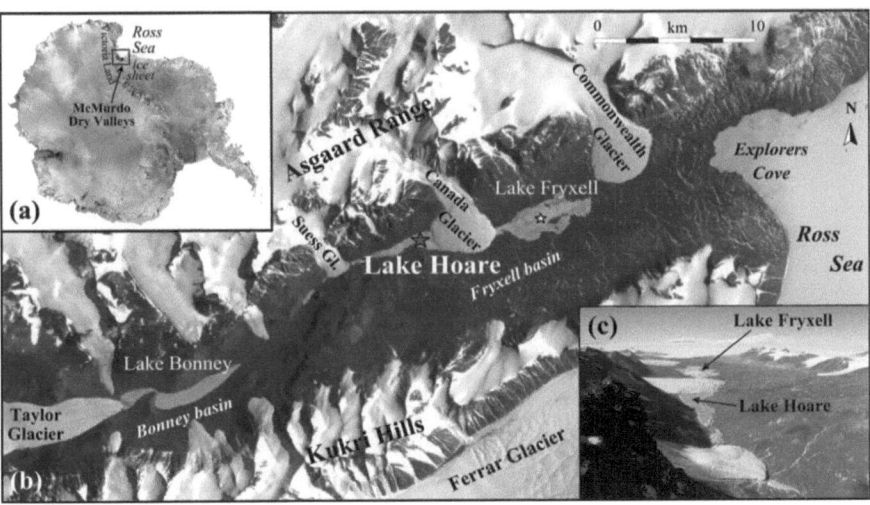

Figure 9. Study site: (a) Location of McMurdo Dry Valleys in Antarctica (U.S. Geological Survey, 2006). (b) Landsat 7 satellite image of Taylor Valley, southern Victoria Land, Antarctica, indicating the location of the most important lakes and glaciers (U.S. Geological Survey, 2007). The red star in Lake Hoare indicates the coring location Lz1020. (c) Photograph of eastern Taylor Valley (source: Gunn, 2006a).

Water temperature below the ice cover is about 0°C, in the deeper parts it can rise up to 1°C (Wharton et al., 1989) (Figure 10b). The main source of water is meltwater of Canada Glacier; temporary streams, like the Andersen Creek, also deliver meltwater from snow or snowbeds to the lake. During lake-level highstands, an additional water source is Lake Chad, a smaller lake located west of Lake Hoare (McKnight and Andrews, 1993). Lake Hoare has no outflow. Thus, lake-level changes are controlled by climatic variations. Loss of lake water is mainly restricted to ablation and sublimation at the ice surface or to evaporation from the moat (Wharton et al., 1989; Wharton et al., 1992).

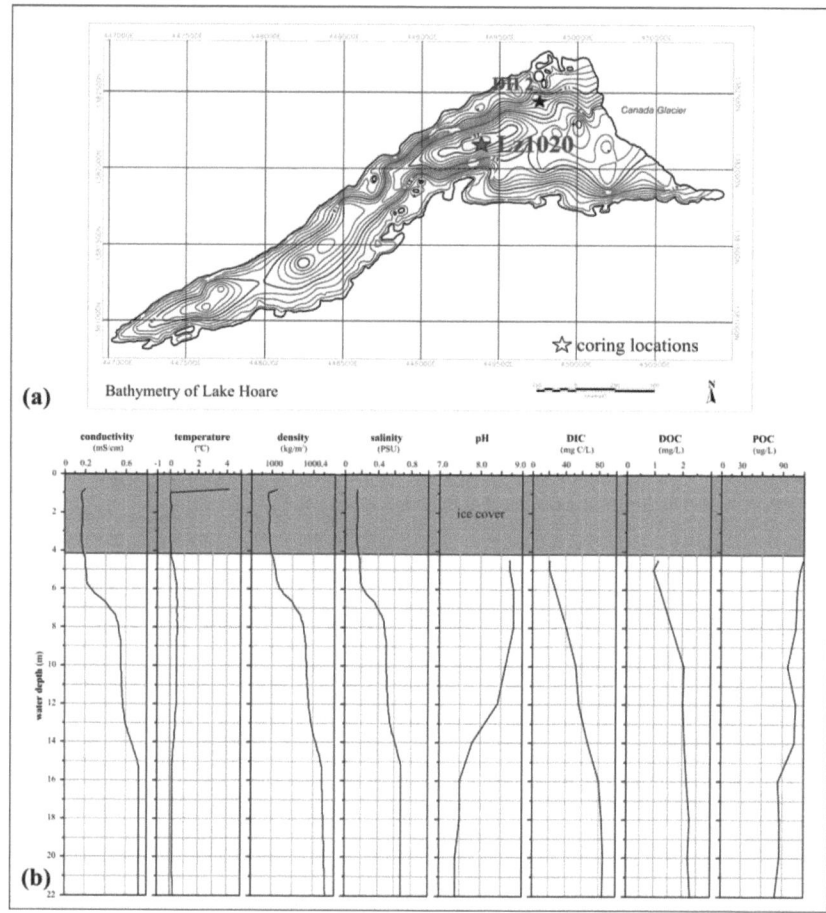

Figure 10. (a) Lake Hoare bathymetry (Schmok and Waddington, 1996) and (b) waterprofile (McMurdo LTER data, season 2002/2003; Lyons and Welch, 2007). Stars in (a) indicate locations of core Lz1020, investigated in this study, and of dive hole 2 (DH 2), mentioned in the study of Doran et al. (1999). DIC – dissolved inorganic carbon, DOC – dissolved organic carbon; POC – particulate organic carbon.

The water of Lake Hoare is characterized by low ion concentrations throughout the water column (Figure 10b). Saline bottom waters, which has been observed in other lakes in Taylor Valley (Lawson et al., 2004) are absent, likely because Lake Hoare is supposed to have been completely desiccated in the past, followed by refilling of the basin with freshwater (Lyons et al., 1998). An oxycline at ~28 m depth separates an anoxic bottom zone from the upper oxic water (Spaulding et al., 1997; Lyons et al., 2000). The perennial ice cover prevents wind-driven induced mixing of the water column and the exchange of gases with the atmosphere (Wharton et al., 1986). Nevertheless, studies of Tyler et al. (1998) reveal a large-scale vertical mixing of the water column, with mixing times of 20-30 years.

4.3 Material and methods

The investigated sediment sequence (Lz1020) was recovered from Lake Hoare during an expedition to Taylor Valley in austral summer 2002/03 (Wagner, 2003). At the coring location (77°37.726'S; 162°52.934'E), a water depth of 32.6 m was measured. A gravity corer (UWITEC Corp.) was used to obtain an undisturbed surface sediment core of 10 cm in length (Lz1020-1). A piston corer (UWITEC Corp.) was used to recover three sediment cores, Lz1020-2 (10-182 cm), Lz1020-3 (0-204 cm) and Lz1020-4 (0-233 cm), which positions were only few meters apart from each other. Immediately after recovering, cores were frozen in a vertical position and later cut into 1 m long sections.

After shipping the cores to the laboratory, all cores were divided lengthwise, described, and photographically documented. Further analyses were only performed on one half of the gravity core Lz1020-1 and the piston core Lz1020-2, as well as on one quarter of the piston core Lz1020-4, the core with maximum penetration. Analyses on the cores were carried out according to Figure 11. Subsampling was performed in 1 cm intervals on the gravity core and in 2 cm steps on the piston cores. After freeze-drying, aliquots of the subsamples were pretreated with 10% H_2O_2 to remove organic matter. Subsequently, samples were sieved to separate the <1 mm grain-size fraction for analyses on a laser particle size analyzer 1180 (CILAS Corp.).

The determination of the amount of volcanic glass was carried out using the heavy mineral fraction. After isolating the fine-sand fraction (125-250 µm) by sieving, the heavy minerals in this fraction were separated by using a centrifuge and a sodium-metatungstate solution (density of 2.83 g/cm^3) as a heavy liquid. Separated heavy minerals were fixed with meltmount on glass slides. Counting of the volcanic glasses was carried out under a polarising microscope, with a minimum of 300 grains counted for statistical significance.

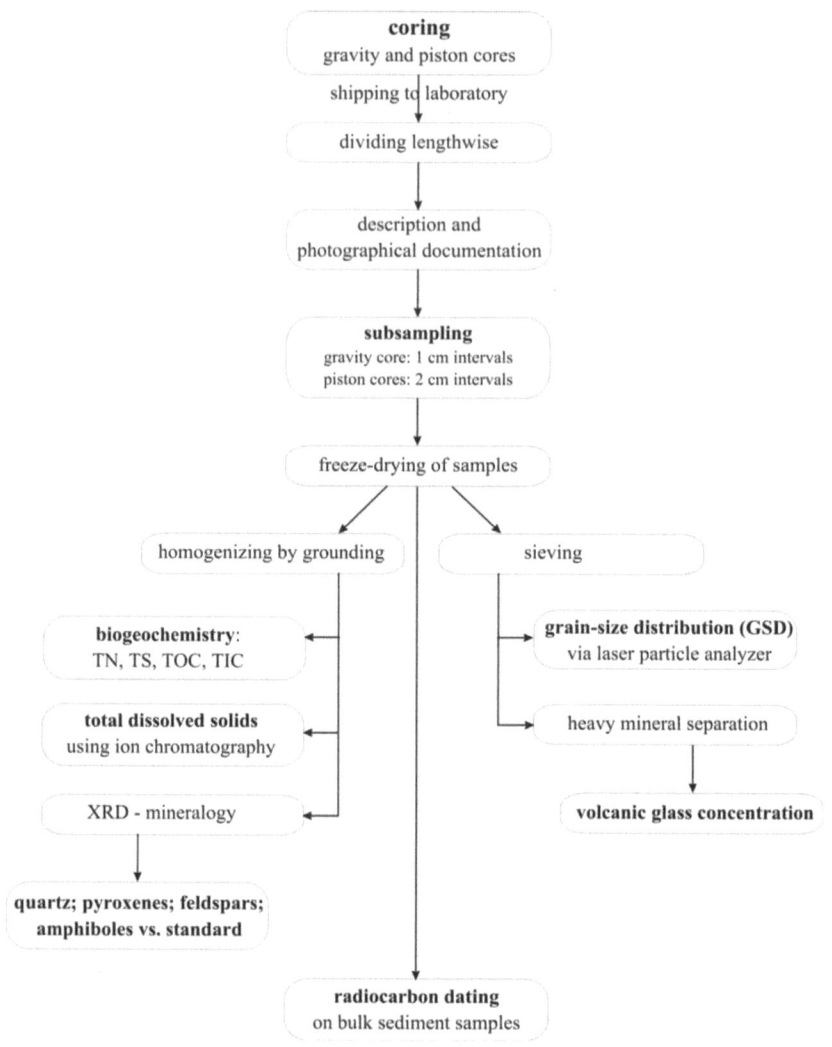

Figure 11. Flowchart of methods for analyzing core Lz1020 from Lake Hoare.

For biogeochemical and mineralogical analyses, an aliquot of each subsample was ground to <63 µm and homogenized. Total carbon (TC), total nitrogen (TN), and total sulfur (TS) contents were measured with an Elementar III (VARIO Corp.) analyzer. Total organic carbon (TOC) content was determined with a Metalyt CS 1000S (ELTRA Corp.) analyzer, after pretreating the samples with 10% HCl at 80°C to remove carbonates. Total inorganic carbon (TIC) content was calculated by the difference of TC and TOC.

Mineralogical analyses were conducted with a miniflex X-ray diffractometer (Rigaku Corp.) with CoKα radiation (30 kV, 15 mA). For a semiquantitative estimation of the mineral content, aliquots of freeze-dried and grounded bulk sediment samples were mixed with an internal standard of Corundum (Al_2O_3) at a sample/standard ratio of 5:1. Random powder mounts were X-rayed from 3 to 40° 2Θ with a step size of 0.02° 2Θ and a measuring time of 2 s/step. The diffractograms were evaluated by using the software "MacDiff 4.2.5" (Petschick, 2001) and following the method described by Neumann and Ehrmann (2001).

Further analyses of the sediment pore water on core Lz1020-4 were conducted by master student Andy Burkemper at the University of Illinois, Chicago (Burkemper, 2007). In order to obtain the salinity of the sediment pore water, it was first necessary to re-dissolve the salts left in the sediment upon freeze drying. Aliquots of dried sediment (core Lz1020-4) at 2 cm resolution were weighed to 0.5000 g and placed into capped centrifuge vials. Ten milliliters of deionized (DI) water was then added to each vial. The cation and anion abundances of the samples, analyzed with an Ion Chromatograph DX-120 (Dionex Corp.) at the Byrd Polar Research Center at Ohio State University, were combined to produce a value of total dissolved solids (TDS). Since the TDS values were not representative of the actual pore water salinity, a correction was necessary in order to apply the measured TDS of the 10 ml solution to the volume of pore water originally present in each sample. The water content data from each sample was used to reconstruct the volume of water that would have been present in each 0.5 g aliquot prior to freeze-drying (for complete details see Burkemper, 2007).

Radiocarbon dating was conducted by accelerator mass spectrometry (AMS) at both the Leibniz Laboratory for Radiometric Dating and Isotope Research in Kiel, Germany (sample numbers beginning with KIA) and University of Arizona at Tucson, U.S.A (sample numbers beginning with AA). The ages of most samples were obtained by dating the humic acid free fraction (HAF) and the humic acid fraction (HA) (Table 4). HAF was created by a standard AAA (acid-alkali-acid) treatment (Grootes et al., 2004). As the TOC contents of the samples, excluding the near-surface sediment sample, were very low with <1 %, high amounts of the sample material were necessary for obtaining a sufficient amount of carbon. The recent reservoir effect in Lake Hoare was estimated by dating the near-surface sediment. With regard to the chronology of the core, the yielded radiocarbon dates were used as rough benchmarks for the chronological constraining of the differentiated sediment units.

Table 4. Radiocarbon ages of the humic acid free fraction (HAF), and the humic acid fraction (HA) from bulk sediment samples of core Lz1020, Lake Hoare.

Sample	Core	Depth (cm)	Material	C (mg)	$\delta^{13}C$ (‰)	^{14}C age (^{14}C years BP)
KIA 30594	Lz1020-1	0-1	HAF	0.5	-26.92	4410 ± 55
KIA 29472	Lz1020-1	1-2	HAF	1.5	-30.05	6210 ± 35
KIA 29473	Lz1020-1	6-7	HAF	0.8	-13.64	5745 ± 45
			HA	1.3	-13.41	5915 ± 40
KIA 30595	Lz1020-4	20-22	HAF	1.6	-24.38	14440 ± 80
			HA	0.8	-20.55	9890 ± 70
KIA 29474	Lz1020-4	50-52	HAF	0.4	-26.11	14580 ± 210
			HA	0.5	-18.66	15000 ± 200
KIA 30596	Lz1020-4	74-76	HAF	1.4	-21.83	22230 ± 170
			HA	0.7	-27.89	18300 ± 190
KIA 29475	Lz1020-4	110-112	HAF	0.3	-26.67	10520 ± 150
			HAF	0.4	-23.17	9370 ± 110
			HA	0.4	-23.01	18680 ± 340
KIA 30597	Lz1020-4	134-136	HAF	0.9	-21.51	19900 ± 180
			HA	0.5	-23.36	18120 ± 250
KIA 29476	Lz1020-4	170-172	HAF	1.1	-17.57	21180 ± 180
			HA	0.5	-23.24	21040 ± 530
KIA 29477	Lz1020-4	188-190	HAF	1.2	-21.60	21210 ± 180
			HA	0.9	-24.30	22530 ± 260
KIA 30598	Lz1020-4	210-212	HAF	1.5	-22.31	20880 ± 160
			HA	0.8	-20.03	20810 ± 250
AA74017	Lz1020-2	10-12	HAF		-23.4	9405 ± 47
AA74018	Lz1020-2	68-70	HAF		-26.2	14577 ± 94
AA74019	Lz1020-2	128-130	HAF		-24.2	10330 ± 47
AA74020	Lz1020-2	166-168	HAF		-21.4	19926 ± 96
AA74021	Lz1020-2	180-182	HAF		-23.5	15759 ± 91
AA71337	LHBC02*	~0	HAF		-27.4	4039 ± 53

*surface mat sample of box core (32 m water depth)

4.4 Results and discussion

4.4.1 Lithology

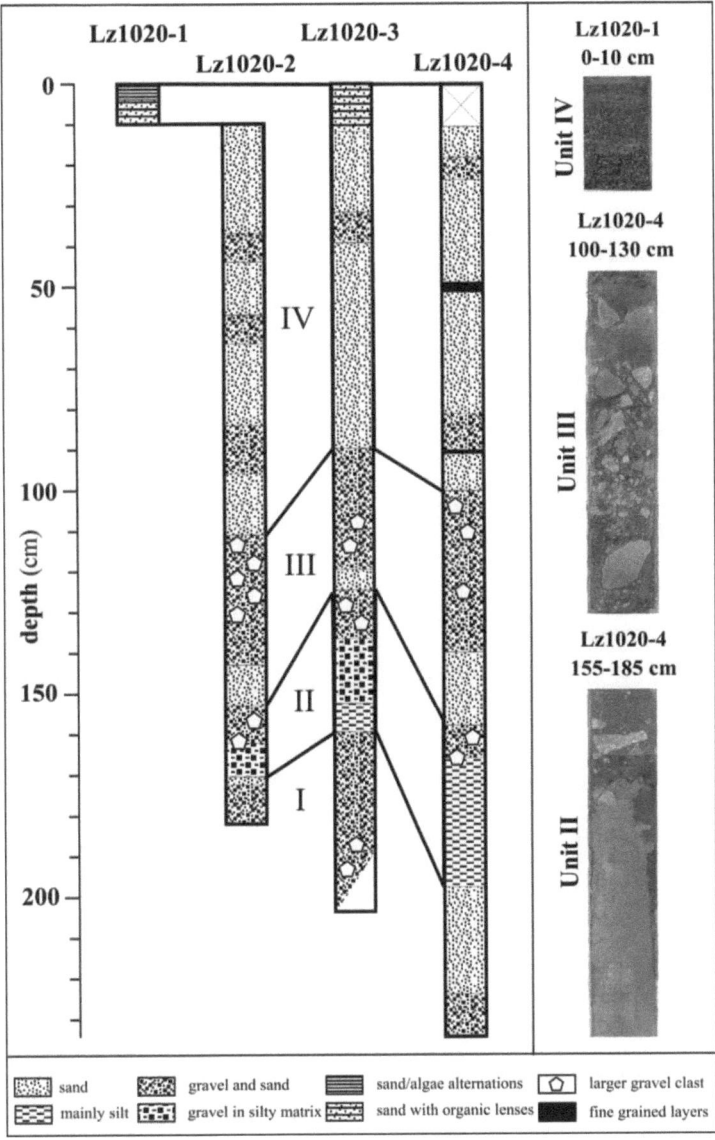

Figure 12. Lithology of Lake Hoare cores, with the core correlation on the left and photos of significant core segments on the right.

4.4.1.1 Core description

Core Lz1020-1

Core Lz1020-1 is the only gravity core recovered from Lake Hoare with a penetration of 10 cm (Figure 12). The base of the core consists of coarse sand and minor amounts of small gravels. At ca. 7 cm depth, the color changes from brownish to grayish, and a small layer of organic material can be found. Between 6 and 3 cm depth, the material mainly consists of medium to coarse sand with very thin gray-greenish organic layers, occurring mostly in the upper part. In the uppermost 3 cm of the core, two distinct horizons of microbial mats, which have thicknesses of up to 5 mm, are separated by a coarse sand layer. The upper one forms the sediment surface. The microbial mats consist of organic rich fine-grained material and contain chrysophyceae cysts and diatoms, such as *Luticola muticopsis* and *Pinnularia cymatopleura* (Figure 13), which has been determined by Dr. H. Cremer (Geological Survey of the Netherlands) and have already been described in more detail by Kellogg et al. (1980), Wharton et al. (1983), and Spaulding et al. (1997). On closer examination, the second microbial layer consists of several fine layers that are separated by thin horizons of sand.

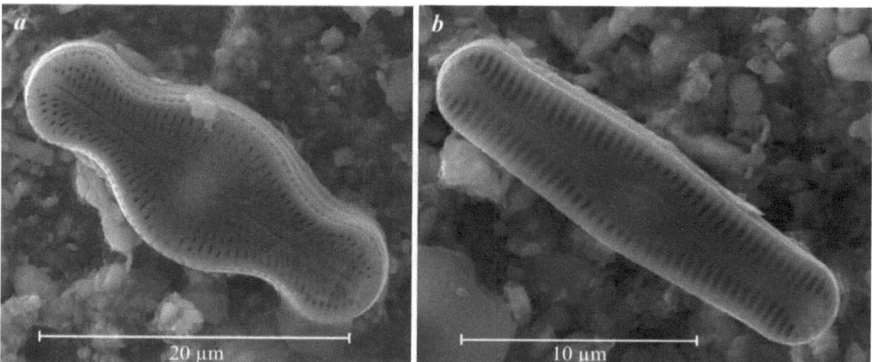

Figure 13. SEM photographs of diatoms in the microbial mats of the surface sediment from Lake Hoare: a) *Luticola muticopsis*; b) *Pinnularia cymatopleura* (determination by Dr. Holger Cremer).

Core Lz1020-2

The first piston core covers a depth of 10-182 cm. At the bottom, from 182 to 172 cm depth, the core consists of brownish-gray gravel and sand with small silty lenses (Figure 12). Above a small layer of medium sand, silty material occurs, which contains larger, poorly rounded clasts. The amount of the clasts increases upward, finally building up a 10 cm thick unit between 165 and 155 cm depth, where clasts of up to 4.5 cm in diameter can occur in a coarse sand matrix. From 155 to 145 cm depth, the core is mainly composed of medium sand.

Upward from 145 cm depth, the grain size increases and turns over in another gravel unit reaching up to 110 cm depth (Figure 12). The poorly rounded clasts have a diverse, but mainly metamorphic lithology, and are coated with silty material. At the top of this section, the amount of medium sand in the matrix increases. Above 110 cm depth, the core mainly consists of these fine to coarse-grained sands. At various depths, e.g. at ca. 90, 60 and 40 cm depth, the grain size increases and smaller gravels occur (Figure 12). Changes in color are also notable, mainly associated with changes in the mineralogy of the grains. The top 25 cm of the core is dominated by grayish mainly medium-sized sand.

Core Lz1020-3

Piston core Lz1020-3 has a sediment depth of 204 cm (Figure 12). The sediments at the core base were disturbed during defrosting of the core. These basal sediments consist of gravel with a fine to coarse-grained sandy matrix. Poorly rounded clasts, which were observed between the base and 180 cm depth, are fragments of volcanic and metamorphic rocks with up to 3 cm in diameter. From 180 to 160 cm depth, the sediment mainly consists of dark gray coarse-grained sand that includes small gravels. Above this section, grain size changes and turns over into mud with a sandy component. From ca. 150 to 140 cm depth larger gravel clasts are embedded in this mainly silty matrix (Figure 12). The diameter of the clasts increases upward up to 3 cm in diameter. The gravel occurs further up, but a change of the matrix to sandy can be observed. From 125 to 120 cm depth, a layer with a dominance of grayish coarse sand marks the transition to another sediment unit, where large volcanic lithics up to 5 cm in diameter occur in a fine-grained sandy matrix. In the upward following section from 110 to 90 cm depth, coarse sand and smaller gravels compose the sediment (Figure 12). The main part of the upper meter of the core consists of grayish-brownish sand. The grain size varies from fine to coarse sand with the latter occurring as lighter colored lenses. From 45 to 35 cm depth, a layer with increased amounts of smaller gravels interrupts the sandy unit. The uppermost 10 cm of the core are characterized by sand that contains organic horizons (Figure 12). The organic material mainly appears as black to gray lenses or as greenish-gray layers, such as those observed in the gravity core. The sands in between mostly show an upward graduation from coarse to fine sand.

Core Lz1020-4

Core Lz1020-4 with a length of 0-233 cm (Figure 12) is the longest piston core recovered from Lake Hoare. The base of the core (233-225 cm depth) is characterized by a layer of clasts of up to 3 cm in diameter. These clasts are coated by silty material and bedded in coarse sand. Above 225 cm depth, brownish-grayish coarse sand contains small lenses of silt, which show an upward increase within this unit. This fining upward sequence continues from 200 to 170 cm depth, forming a silty unit, where only small amounts of medium sand occur. At the top of this section (170-160 cm depth), larger clasts are embedded in the fine-grained material (Figure 12). The size of the poorly rounded clasts increases upward up to 4 cm in diameter. The matrix changes from being silty to sandy. From 160 to 140 cm depth, the sediment is firstly dominated by small gravels and coarse sand, and further on by medium sized sand. At 140 cm depth, smaller gravels in a coarse sand matrix mark the beginning of another clastic unit (Figure 12). The size of the poorly rounded clasts increases until the core depth of 100 cm. Furthermore, the top of the clastic unit shows increased amounts of medium sand. From 100 to 80 cm depth, the core mainly consists of medium to coarse sand and smaller gravels. The upper 80 cm of the core is characterized by a dominance of sandy material (Figure 12). The sporadic occurrences of smaller clasts, especially at ~20 cm depth, and a clayey-silty layer of about 2 cm thickness at ca. 50 cm depth interrupt the sandy sedimentation. In some parts, color changes can be observed, whereas finer sands are dark colored, coarse sands appear light colored mostly due to their mineralogy. The top of the core (20-10 cm depth) consists of grayish medium-sized sand, which seems to contain organic material. The material of the uppermost 10 cm is lost due to defrosting of the core (Figure 12).

4.4.1.2 Core correlation and discussion

The cores recovered from Lake Hoare show a similar lithology, dominated by coarse-grained clastic matter. The sediment sequences all contain the same main units, but in differing thicknesses. Based on their lithology, four units can be defined (Figure 12).

Unit I

Although reaching different sediment depths, the bases of all cores are characterized by gravelly and sandy sediments with poorly rounded clasts (Figure 12). The thickness of this section varies remarkably between 10 and 40 cm. In core Lz1020-4, the upper part of this unit is characterized by a dominance of sandy material.

The large clasts of unit I seem to have a glacial origin, but the matrix, which makes up the main part of the accumulated sediment, is sandy and silty suggesting deposition in a lake system. Hendy (2000b) suggests that clastic grains with a diameter larger than 1 cm cannot melt their way through the perennial ice cover of today dry valley lakes. According to this, the occurrence of gravel in the sediments is indicative for a non-perennial ice cover.

Alternatively, Hendy et al. (2000) and Hall et al. (2006a), to explain the presence of coarse grained sediments, suggest that an internal lake-ice conveyor belt forms in proglacial lakes.

Coarse till is transported on the lake ice to the edge of the lake, where it is dumped in the moat zone. Hall et al. (2000) describe ridges of the present valley slopes, that contain such high amounts of drift sediments, consisting of coarse sand up to boulders that accumulated in moats of former lakes. These layers can be used as indicators for former lake levels. In connection with low lake levels, gravel-rich deposits can also originate from slope fall. Hendy et al. (2000) also suggest that local accumulations of gravel could be due to dropstone release from icebergs, which are drifting as part of the ice cover through proglacial lakes. However, since the gravelly sands also contain fine-grained material, deposition seems to have taken place in a lacustrine environment. The occurrence of gravel could be traced back to the existence of a non-perennial ice cover.

Unit II

Unit II occurs between 200 and 130 cm depth and consists of fine-grained material with larger clasts in the upper part (Figure 12). It reaches a maximum thickness of 40 cm in core Lz1020-4. At the top of unit II larger poorly rounded gravel clasts are lying in the fine-grained matrix like dropstones. Unit II is not as pronounced and thick in cores Lz1020-2 and Lz1020-3 as in Lz1020-4. However, the silty composition and the succession of sediments suggest being the same unit. The mixture of the clasts and the silty material in cores Lz1020-2 and Lz1020-3 may result from disturbances during the coring process.

The silty sediment composition is characteristic for large proglacial lakes (Hall et al., 2006a). And since the surrounding of the present Lake Hoare is characterized by sandy sediments, the mainly silty material indicates a high water column, where maybe a lake-ice conveyor belt of a proglacial has operated. According to Hall et al. (2000), unit II could be identified as the glaciolacustrine silt facies. This would suggest that the coring site was occupied by Glacial Lake Washburn. The clasts in the upper part could be explained as dropstones that fell into the fine-grained material.

Unit III

A significant change occurs between 160 and 100 cm depth in form of a 40 to 60 cm thick gravelly section (unit III in Figure 12). The most prominent characteristics of this unit are the very large, poorly rounded clasts with diameters of up to 5 cm. Following the lake-ice conveyor belt model by Hendy et al. (2000) and Hall et al. (2006a), this unit could be either moat zone deposits in a proglacial lake or local accumulations of gravel due to dropstone release from icebergs. But with regard to the thickness of this layer, iceberg release seems to be very unlikely.

Since fine material is mostly lacking, relatively high transport energies during or after deposition of this horizon are indicated. So it is more likely that these sediments are lag deposits. The occurrence of bigger clasts in the upper part of this horizon is comparable to recent subaerial stone pavements in the surrounding of the lake (McKnight et al., 1999) and suggests similar formation conditions in earlier times, with deflation of the finer sediments during drawdown of the lake. The lithological composition of unit III may indicate subaerial conditions at the coring site, probably as a result of lake level lowering.

Unit IV

In the upper part of the cores, above ca. 100 cm depth, the sediments mainly consist of fine to coarse-grained sand, with some interspersed layers of gravel or clayey silt (unit IV in Figure 12), which indicates relatively stable sedimentation conditions over a longer period.

The generally coarse grained sediments in Lake Hoare can be explained by the catchment of the lake. Today Lake Hoare is located in a narrow section of Taylor Valley, where material from the valley sides is transported to the lake on a short distance. Additionally, Canada Glacier blocks the down-valley blowing catabatic winds, resulting in the accumulation of a high amount of wind-blown sediments on the rough ice cover of Lake Hoare. These sediments can migrate by freezing and thawing processes through the ice cover or can fall through cracks in the ice cover and eventually accumulate on the lake bottom, forming sand mounts and ridges (Nedell et al., 1987; Squyres et al., 1991; Andersen et al., 1993). Hendy (2000b) described a sorting of the grain sizes during this process, with sandy and silty material passing through the ice cover and gravel and bigger clasts remaining on the ice surface.

This regime of episodic sediment deposition leads to a highly locally and temporally variable accumulation on the lake bottom. Sediment structures observable in the sandy unit IV of the cores (Figure 12), which consists of a succession with coarser light coloured sand at the bottom and finer dark colored sand at the top, could be explained by typical depositional processes in Lake Hoare described by Squyres et al. (1991).

The overall negligible amounts of fine-grained material are likely due to the low number of inflowing streams, which supply glacial meltwater and fine-grained suspension load. Only when melting through the ice cover is reduced, predominately suspended particles in the water column can be deposited, explaining thin fine-grained horizons in the sandy sediments.

A distinct change in sediment composition is observable in the gravity core Lz1020-1 containing an undisturbed sample of the sediment surface. In the uppermost 10 cm, horizons of microbial mats of up to a few millimeters thickness alternate with coarse-grained sand layers of up to a few centimeters thickness (Figure 12). Indications for organic sedimentation can also be found in core Lz1020-3 in form of organic lenses preserved in the sand. Previous studies on surface sediments of Lake Hoare (Spaulding et al., 1997; Doran et al., 1999) also describe these alternating layers of microbial mats and coarse-grained sediments. Whilst the microbial mats contain a higher percentage of fine material and are formed, when sediment accumulation rates are low, the interspersed sandy horizons represent higher sedimentation rates and probably single events, such as the formation of sand mounds and ridges (Squyres et al., 1991; Andersen et al., 1993).

4.4.2 Biogeochemistry

Total organic carbon (TOC), total inorganic carbon (TIC), total nitrogen (TN), and total sulfur (TS) contents are very low throughout most of the sediment sequence of cores Lz1020-1 and Lz1020-4 (Figure 14), being close to their limits of detection. Low contents of these biogeochemical parameters are common in Antarctic lake sediments, since input of terrestrial biomass is lacking or negligible, and since biomass in lakes is primarily produced by microbes, which are often decomposed after sedimentation (Fountain et al., 1999; Hodgson et al., 2004). Therefore, elevated amounts of TOC, TN and TS are only found in the uppermost part of the core (unit IV), where microbial mats are preserved (Figure 14).

The TS content displays a gradual increase in the upper 20 cm of the sediment sequence, interrupted by low TS levels at ~10 cm depth (Figure 14). The TS spike at ca. 15 cm depth shows no correlation with enhanced accumulation of organic matter, so its deposition probably indicates the formation of pyrite in anoxic water (Håkanson and Jansson, 1983). In contrast, the maximum TS content, near the sediment surface, is correlated with an increase in TOC and TN contents (Figure 14), which reflects the presence of the microbial mats. Thus, this upper TS maximum can be due to the presence of both organic sulfur and pyrite, which is present in near-surface sediments deposited in the deep, anoxic pockets of Lake Hoare (Bishop et al., 2001).

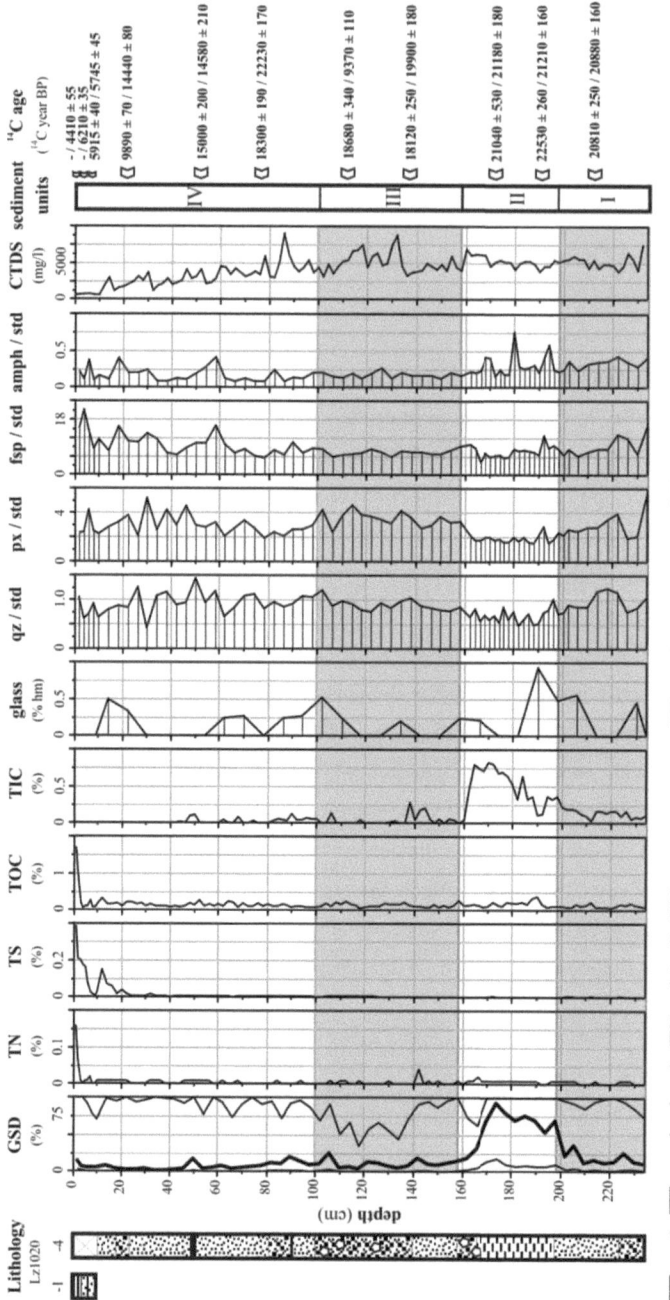

Figure 14. Characterization of core Lz1020: Lithology, grain-size distribution (GSD, with cumulative clay, silt, sand, and gravel fraction from left to right), total nitrogen (TN), total sulfur (TS), total organic carbon (TOC), total inorganic carbon (TIC), concentration of volcanic glass in the 125-250 µm heavy mineral fraction, amounts of quartz (qz), feldspars (fsp), pyroxenes (px), and amphiboles (amph) versus standard (std), corrected total dissolved solids (CTDS), and dated horizons with their ^{14}C ages (^{14}C years BP) for the humic acid (HA) and humic acid free (HAF) fraction of core Lz1020 from Lake Hoare. The units were differentiated due to their lithological properties.

The TIC content most likely reflects the presence of carbonates in the form of calcite or aragonite (Figure 15). The highest values occur in unit II with a maximum of ~0.8 % at 165 cm depth, which is followed by a significant drop (Figure 14). The observed increase in unit II suggests that TIC was formed by carbonate precipitation during a period of enhanced evaporation and lake drawdown (Lawrence and Hendy, 1989; Hendy, 2000a), when ion concentrations in the lake increased. Similarly, the elevated amounts of TIC in the lower part of unit III at ca. 140 cm depth could be traced back to a falling lake level due to evaporation. The small fluctuations of the TIC values between 130 and 40 cm are probably related to lake level oscillations (Figure 14). Since the TIC contents in the upper 40 cm of the sediment sequence are below the detection limit, ion concentrations in the lake water likely were not sufficient to induce carbonate precipitation. Carbonate and aragonite horizons have also been observed in sediment sequences from Lake Fryxell and were interpreted as a result of increased biogenic productivity, enhanced evaporation, and lake level lowering (Wharton et al., 1982; Lawrence and Hendy, 1989; Wagner et al., 2006). Increased productivity, as it would be indicated in a higher content of TOC or TN can, however, not be observed in the sediment sequence from Lake Hoare (Figure 14).

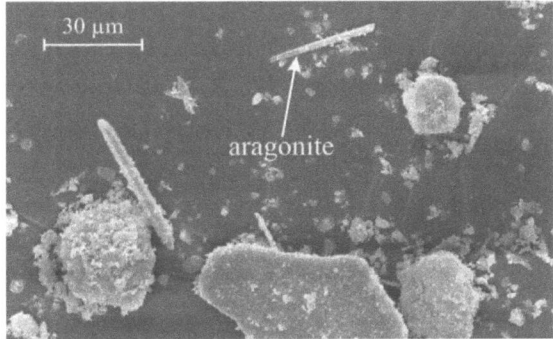

Figure 15. SEM photograph of aragonite needles in core Lz1020-4, 170-172 cm depth, from Lake Hoare.

In the lower part of the core, the values for 'Corrected total dissolved solids' (CTDS in Figure 14) are relatively constant around 5000 mg/l with a slightly increasing trend towards the top of unit II. In lower unit III, a decrease in the CTDS values can be observed, which is followed by a positive shift within the gravel section between 130 and 110 cm depth. A second peak occurs in unit IV at 90-80 cm depth (Figure 14). Due to the fact, that higher values can be observed subsequent to TIC peaks, these sediments could be deposited in a lake with a low level and saline waters. From about 80 cm depth to the top of the core, there is a steady decrease in salinity, which could be interpreted as an increasing lake level. The CTDS values for the uppermost samples in the core match well with the current value for Lake Hoare bottom water of ~600 mg/l (Lyons and Welch, 2007). Compared to Lake Fryxell, where the bottom waters show TDS values of ~6000 mg/l (Lyons and Welch, 2007), the pore water in the uppermost part of the Lake Hoare core as well as the current bottom waters are tenfold lower. A possible explanation could be the refilling of the basin with glacial meltwater. In respect to the environmental conditions, Canada Glacier advance is responsible for the damming of Lake Hoare. The decreasing salinity in the upper 80 cm of the core (Figure 14) would suggest a significant influence of the glacier by delivering fresh meltwater to the lake during this time.

4.4.3 Mineralogy

Source areas of the sediments can be deduced from their mineralogical composition. X-ray diffractometry (XRD) provides semiquantitative data of the major mineralogical components of the sediment. By the evaluation of the diffractograms, the abundance of the main mineral groups quartz, feldspars, pyroxenes and amphiboles can be estimated.

Mineralogical composition of the cores is representative of almost all local rock types and is also present in the Quaternary substrate that mantles the valley floors and is the main source for clastic input to the lakes. The data show relatively small variations in the composition of the Lake Hoare sediments throughout the core (Figure 14). The contents of quartz and pyroxenes show a similar behavior with higher amounts at the base in units I and III. Lowest contents can be observed in unit II. Maximum contents occur in the upper part of unit IV, with a pyroxenes/standard ratio of ~5 at 30 cm depth, and a quartz/standard ratio of ~1.5 at 50 cm depth (Figure 14).

Feldspar is the most abundant mineral, showing the highest ratios of the mineral groups (Figure 14). This can be explained by the crystalline basement exposed in the vicinity of Lake Hoare (Porter and Beget, 1981). In unit VI above 60 cm depth, the ratio increases and reaches a maximum of ~18 at 4 cm depth (Figure 14). Since feldspar may originate from granitoid sources, the elevated amounts may be traced back to an increased influence of Canada Glacier, which probably delivers material eroded from the Granite Harbor Intrusive Complex directly to Lake Hoare.

The amphibole contents are low. Increased amounts occur in the lower part of the core, in units I and II, where a maximum of the amphibole/standard ratio of ~0.6 is reached. The upper part of the core shows an overall low amphiboles/standard ratio being only occasionally higher than 0.25 (Figure 14).

Similarly, the occurrence of volcanic glass is very low throughout the Lake Hoare sequence. Some fluctuations occur, with a maximum at 190 cm depth in unit II, but overall values are less than 1 % of the heavy mineral fraction (Figure 14). The amount of volcanic glass is thus significantly lower than in the sediments of Lake Fryxell, where it reached up to 20 % of the heavy mineral fraction (Wagner et al., 2006). The increased occurrence of volcanic glass in lake sediments, e.g. in Lake Fryxell, deposited during the Weichselian was attributed to an advanced Ross Ice Shelf, which delivered relatively large amounts of volcanic glass from the McMurdo Volcanic Group on Ross Island to the dry valleys (Denton and Hughes, 2000; Denton and Marchant, 2000; Hall and Denton, 2000b). The overall low amount of volcanic glass in Lake Hoare sediments implies that sediment supply from an advanced Ross Ice Shelf was not so significant at this site, probably due to the more distal position from the ice margin. Another explanation could be that the glass originates from smaller, local sources in upper Taylor Valley (Angino et al., 1962). Otherwise, the elevated amounts of amphiboles in the lower core may also originate from rocks of the McMurdo Volcanic Group (Ehrmann and Polozek, 1999; Neumann and Ehrmann, 2001).

In summary, the elevated volcanic glass concentrations and amphibole contents in units I and II highly indicate a supply from the McMurdo Volcanic Group, from which the influence of an advanced Ross Sea ice sheet at the mouth of Taylor Valley can be deduced. The dominance of quartz, feldspar, and pyroxenes reflect the prominence of the bedrock in the vicinity of Lake Hoare as source area. Changes in the amounts of quartz, feldspar, and pyroxenes may rather indicate changes in the depositional environment and transport, as observable by an increasing feldspar input in unit IV probably being induced by an advance of Canada Glacier.

4.4.4 Chronology

4.4.4.1 Radiocarbon ages

Radiocarbon dating is a common method for dating lacustrine sediments from Antarctica (Doran et al., 1999). However, different sources of carbon bias the date of sediment accumulation, such as input of old carbon from glacier meltwater or erosion of older lacustrine sediments in the catchment. In addition, the perennial ice cover on most lakes and salinity stratification prohibits the exchange of CO_2 between the atmosphere and the water column. As a result, the radiocarbon ages of sediments are affected by two components: the inherited age and the residence age of the lake water (Hendy and Hall, 2006).

The modern reservoir effect is best assessed by dating surface sediments. In our core, modern microbial mats on top of the core (sample KIA 30594) yielded an age of about 4410 ^{14}C years BP (Table 4). This is older than the surface sediment age of Doran et al. (1999) of ~2500 ^{14}C years BP (Figure 16) in a core recovered in 11 m water depth from Lake Hoare (DH2 in Figure 10a). The significantly older age at the Lz1020 coring site may result from the greater water depth, which is more affected by stratification of the water column (c.f. Hendy and Hall, 2006). This is supported by another date, which was obtained from the surface of a box corer sample. Sample AA 71337 was recovered from 32 m water depth and yielded an age of 4039 ^{14}C years BP (Table 4).

Throughout the core, the HAF ages show no consistent age-depth correlation. However, the ages of core Lz1020-2 fit well to the ages of cores Lz1020-1 and Lz1020-4, except for small depth variations (Figure 16; Table 4). The relatively young age of sample AA 74021 with respect to its depth can be relativized, since the sample originates from unit I (Figure 16).

Challenges for creating a chronology evoke from two reversals occurring in the age-depth correlation (Figure 16) at 110-112 cm depth (KIA 29475) and at 210-212 cm depth (KIA 30598). We explain these reversals through changing carbon reservoir effects in the lake over time. Equilibration of the water column with the atmosphere, e.g. during a complete desiccation of a lake or in times of a low water level and a lacking ice cover, may result in a zeroing of the reservoir effect. We assume that the radiocarbon-dated sample KIA 29475 from 110-112 cm depth (Table 4) provides a "real age" of ~10,000 ^{14}C years BP, because unit III was very likely deposited under subaerial conditions. Even if the yielded ages show a large discrepancy between the HA and HAF fraction, and the organic carbon content of sample KIA 29475 is very low, the repeated dating in the HAF fraction produced nearly the same age (Table 4). This date is also reproducible by sample AA 74019 from unit III in core Lz1020-2, which yielded an age of 10,330 ^{14}C years BP in the HAF fraction.

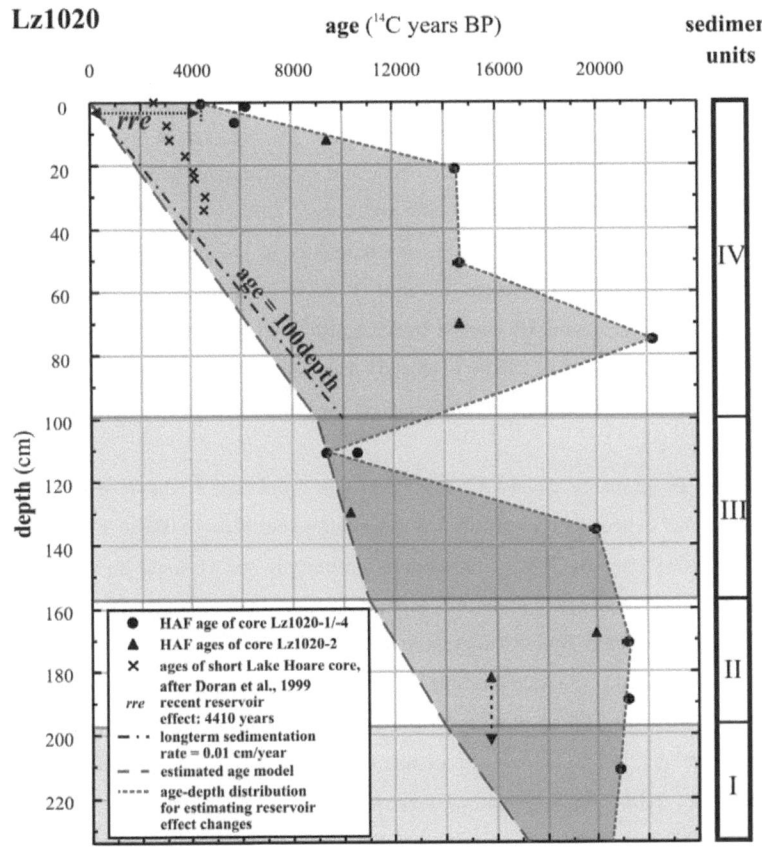

Figure 16. Age-depth distribution of core Lz1020 from Lake Hoare. The figure shows the depth distribution of HAF (humic acid free fraction) ages. The recent reservoir effect (*rre*) in Lake Hoare could be estimated by dating the sediment surface and amounts 4410 years. For the upper part of the core, the results of high-resolution radiocarbon dating on another Lake Hoare core, conducted by Doran et al. (1999), is added to the figure, and the long-term sedimentation rate of ca. 0.01 cm/year (Squyres et al., 1991) is illustrated as the dashed-dotted line. Following the discussion in chapter 4.4, the dashed line represents the estimated age model for core Lz1020. In comparison to the estimated age model, changes in the reservoir effect can be deduced from the depth distribution of the HAF ages (dotted line).

4.4.4.2 Ages of units

Since the ages of the HAF fraction do not show a consistent age-depth correlation for the upper core (unit IV), we used published data to help constrain the relationship. For these reasons, ages of the sediment units can only be regarded as approximated ages and are given in years BP within this study.

Doran et al. (1999) conducted high-resolution radiocarbon dating on microbial mats in a short core recovered from Lake Hoare (Figure 16), and they calculated an average sedimentation rate of ca. 0.015 cm/year. This fits relatively well to the long-term sedimentation rate of ca. 0.01 cm/year, estimated by Squyres et al. (1991). For the chronological constraining of unit IV, we deduced a sedimentation rate of ca. 0.011 cm/year by extrapolating between the top of the core and sample KIA 29475 with an age of 9370 ^{14}C years BP (Figure 16; Table 4), and also in approximation to the long-term sedimentation rate after Squyres et al. (1991). Consequently, unit IV possibly represent approximately the last 9000 years. The age of unit III can be better constrained due to the reproducible age of samples KIA 29475 and AA 74019 (Table 4). Because the sedimentation rate should be higher in gravel deposits, unit III probably covers the period between 11,000 and 9000 years BP (Figure 16).

Since unit IV probably covers the last 9000 years and unit III likely originate from a regional detected evaporation event occurring sometime between 10,000 to 11,000 years ago in the dry valleys (Hendy, 2000a), the underlying units II and I should be Lake Washburn deposits. The sedimentological properties support this assumption. Proglacial Lake Washburn is supposed to have filled Taylor Valley at least from 23,800 ^{14}C years BP (Hall et al., 2000). Samples of units I and II (KIA 29476, 29477, 30598 in Table 4) yielded ages being older than 20,000 ^{14}C years BP. These ages are probably too old because of the reservoir effect contained (see discussion below). In comparison with core Lz1021 from Lake Fryxell (Wagner et al., 2006), the fine-grained material of unit II could coincide with the period between ~14,000 and 11,000 years BP (Figure 18). Unit I at the core base would consequently be older than 14,000 years BP. Assuming a relatively constant sedimentation rate in units I and II, the core base would be ~17,000 years old.

4.4.4.3 Reservoir effect changes

Past changes of the reservoir effect can be estimated by the difference between the age model and the yielded radiocarbon dates (Figure 16). During the last glacial maximum, eastern Taylor Valley was occupied by Lake Washburn. This glacial lake is supposed to have reached lake levels higher than 200 m a.s.l. (Hall et al., 2000) with a discontinuous lowering of its lake level towards the Holocene (Hall and Denton, 2000b; Wagner et al., 2006). During the lake level lowering, lacustrine sediments became subaerially exposed at the slopes and could have been affected by erosion. The mixture of organic carbon from these older lacustrine sediments with in-situ organic material, produced in declining Lake Washburn, forms the bulk organic matter, which is used for radiocarbon dating of Lake Hoare sediments. Additionally, an increased supply of carbonate from the catchment, i.e. old soil carbon, may result in higher reservoir effects. The input of old, glacial DIC derived from the RIS and aging of the lake water due to lacking mixing of the water column may also be responsible for a progressively increasing reservoir effect in former Lake Washburn (Hendy and Hall, 2006), represented in units I and II (Figure 16). A smaller reservoir effect can be observed in the lower core for samples KIA 30598 and AA 74021 (Table 4), which are assigned to unit I (Figure 16). This suggests that during the formation of unit I, erosional processes from the valley slopes played a minor role. This could be the case either during times of lake-level highstands or during dry climatic conditions, when meltwater and material supply to the lakes was restricted. A zeroing of the reservoir effect due to an equilibration with the atmosphere can be assumed in unit III. The partly high discrepancies of the HAF ages to the estimated age model in unit IV (Figure 16) can also be explained by a variable reservoir effect in the past. Some time between 9000 and 5000 years BP, the reservoir effect seems to increase again as a result of the aforementioned reasons to maximum of about 15,000 years in the course of the refilling of Hoare basin. The decreasing reservoir effect in the upper core (Figure 16) could be traced back to a decreasing supply of old carbonate and Lake Washburn sediments, with a contemporaneous elevated stream input, which delivers material equilibrated with the atmosphere.

4.5 Implications for the environmental history of Taylor Valley

Units	Lake Hoare	Lake Fryxell	Environmental history
IV < 9000 years BP towards recent conditions	no volcanic glass ⇨ RIS retreat from valley mouth no TIC ⇨ separation of lake systems by advance of Canada Glacier	no volcanic glass TIC	at least < 2500 years BP 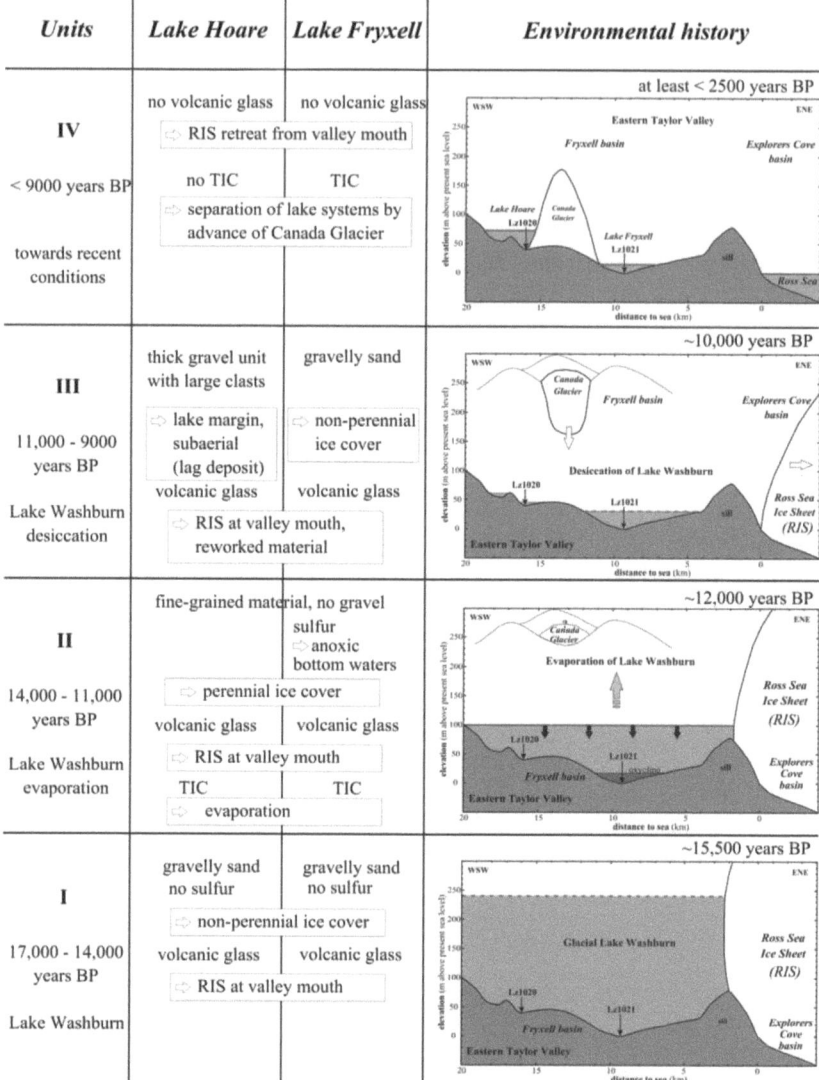
III 11,000 - 9000 years BP Lake Washburn desiccation	thick gravel unit with large clasts ⇨ lake margin, subaerial (lag deposit) volcanic glass ⇨ RIS at valley mouth, reworked material	gravelly sand ⇨ non-perennial ice cover volcanic glass	~10,000 years BP
II 14,000 - 11,000 years BP Lake Washburn evaporation	fine-grained material, no gravel sulfur ⇨ anoxic bottom waters ⇨ perennial ice cover volcanic glass ⇨ RIS at valley mouth TIC ⇨ evaporation	volcanic glass TIC	~12,000 years BP
I 17,000 - 14,000 years BP Lake Washburn	gravelly sand no sulfur ⇨ non-perennial ice cover volcanic glass ⇨ RIS at valley mouth	gravelly sand no sulfur volcanic glass	~15,500 years BP

Figure 17. Reconstruction of the environmental history of eastern Taylor Valley, Antarctica. The figure illustrates stages of the environmental history of eastern Taylor Valley, reconstructed by the lake sediment records of lakes Hoare and Fryxell, and based on the four units of core Lz1020 (first column). Sediment properties of lakes Hoare and Fryxell, given in the second and third column, respectively, are used for deducing paleoenvironmental information (TIC = total inorganic carbon; RIS = Ross Sea Ice Sheet). The fourth column shows scenarios for the profile of eastern Taylor Valley (in fortyfold vertical exaggeration) at particular times within the different stages of lake history.

4.5.1 Unit I (~17,000-14,000 years BP) – Lake Washburn

The sediments of unit I at the core base were deposited during the late Weichselian, probably between 17,000 and 14,000 years BP. At that time, a proglacial lake occupied the lower part of Taylor Valley (Stuiver et al., 1981). Glacial Lake Washburn was dammed by the Ross Sea Ice Sheet (RIS) that blocked the mouth of Taylor Valley (Figure 17), and was mainly fed by meltwater derived from the RIS (Hendy, 2000a). It should be noted that alpine glaciers were in a retreated position throughout the whole late Weichselian (Denton et al., 1989).

From the occurrence of volcanic glass and the elevated amounts of amphiboles in unit I of the Lake Hoare record (Figure 14), it can be concluded that the sediments of unit I were deposited when the RIS was flowing across McMurdo Sound and delivering volcanic material of the McMurdo Volcanic Group from Ross Island to Taylor Valley (Figure 3). From deposits observed in Taylor Valley, it is assumed that the lake had established a mechanism of drift deposition, which can be described as a lake-ice conveyor belt (c.f. Hendy et al., 2000; Hall et al., 2006a). Hall and Denton (2000b) relate deposition of gravelly and sandy material, comparable to the sediments in unit I of core Lz1020 (Figure 14), to periods of an increased activity of the lake ice conveyor of Lake Washburn. Furthermore, these authors suppose that a period with an increased lake-ice conveyor sediment deposition occurred between 16,500 and 15,500 ^{14}C years BP, which nearly coincides with the chronology model for unit I. However, since gravel-rich sediments are mainly ascribed to moat deposits in terms of the lake-ice conveyor belt model (Hendy et al., 2000), and paleodeltas dating from this period do not indicate a lowstand of Lake Washburn (Hall and Denton, 2000b), other mechanisms must have lead to the deposition of unit I. We assume that Lake Washburn had a non-perennial ice cover during this time (Figure 17), since sediments of gravel-size cannot melt through a permanent ice cover (Hendy, 2000b).

In this context, the Taylor Dome ice core record shows indications for a warming from 16,000 to 14,000 years BP (Steig et al., 2000). The warmer temperatures could have supported lake-ice thinning or even a regular thawing of the ice cover. This process must have worked across the whole Lake Washburn, because the Lake Fryxell record (core Lz1021) also shows gravelly and sandy sediments, which are comparable to those of unit I of the Lake Hoare sequence (core Lz1020), during this period (Figure 18).

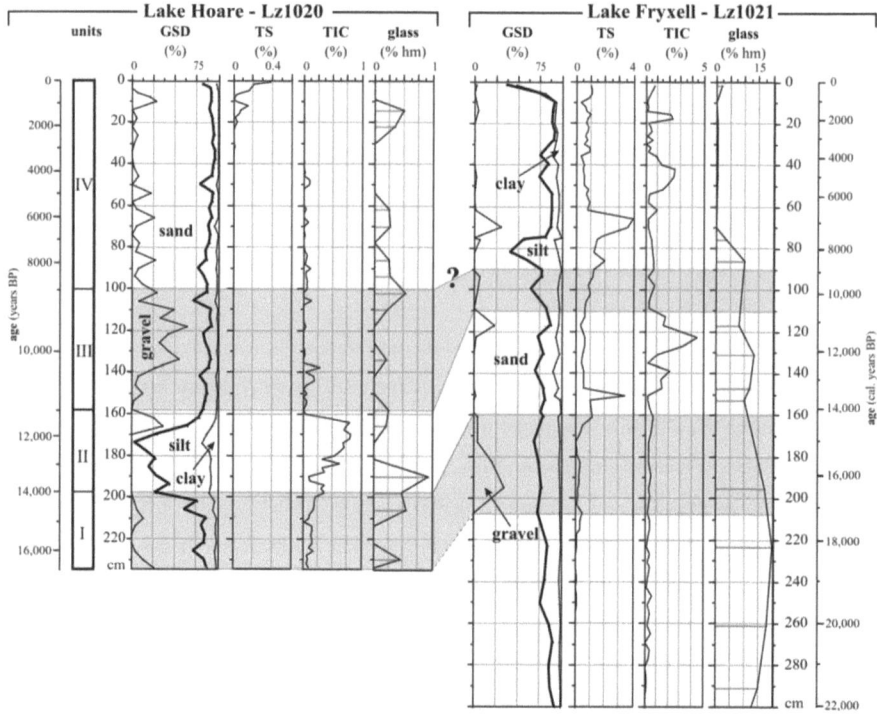

Figure 18. Correlation of core Lz1020 from Lake Hoare and core Lz1021 from Lake Fryxell The figure shows a possible correlation of the records of both lakes, based on four different units of core Lz1020 from Lake Hoare. It illustrates the most important parameters grain-size distribution (GSD, with cumulative gravel, sand, silt, and clay fraction from left to right), total sulfur (TS), total inorganic carbon (TIC), and volcanic glass concentration in the heavy mineral fraction. For chronological constraints, the figure includes the ages for the Lake Fryxell record based on the age model by Wagner et al. (2006), and the ages for the Lake Hoare record based on our estimated age model.

4.5.2 Unit II (~14,000-11,000 years BP) – Lake Washburn evaporation

Unit II is characterized by silty and sandy sediments (Figure 14) deposited in the area of todays Lake Hoare. Volcanic glass and elevated amphibole contents detectable in unit II of core Lz1020 (Figure 14) can be traced back to a still persisting influence of the RIS. The dominance of fine-grained material and the lack of gravels, except for the uppermost part of unit II, may be explained by the existence of a perennial ice cover (Figure 17). Furthermore, the Lake Fryxell record indicates the establishment of anoxic bottom water by a significant TS peak at ca. 150 cm depth (corresponding to ~13,500 cal. years BP) in core Lz1021 (Figure 18), which would support the assumption of a permanent ice cover (Figure 17). This would imply cold climate conditions. Between 14,600 and 12,700 ^{14}C years BP, the RIS advanced into Taylor Valley to its maximum extent of the LGM, which can also be related to a lower lake level of Lake Washburn (Hall and Denton, 2000b; Hall et al., 2000). During the maximum extent of the RIS, very cold and dry climatic conditions may have dominated. A period of aridity in the McMurdo Dry Valleys region between 13,000 and 12,000 years BP, probably correlated to the Younger Dryas stade, can be deduced from the Taylor Dome ice record (Grootes et al., 2001). These climatic conditions may have initiated an evaporation of the large glacial lake. Terrestrial archives like paleodeltas indicate a lowering of the lake level starting at ~13,000 ^{14}C years BP, whereas the Fryxell basin was firstly isolated by ~12,000 ^{14}C years BP and finally by ~11,000 ^{14}C years BP (Hall and Denton, 2000b). The distinct increase in TIC in both Lake Hoare and Lake Fryxell records (Figure 18) can be traced back to a gradual evaporation of the water column and a lowering of the lake level (Figure 17), leading to cumulative concentration of salts and the formation of aragonite or calcite (c.f. Hendy, 2000a; Lawrence and Hendy, 1989).

The coarsening of grain size up to sand and gravel in the uppermost part of unit II (core Lz1020) may indicate lake-ice thinning, probably caused by an increased salinity (Figure 14). Coarse-grained sediments, which had been accumulated on the ice cover in prior times, were released and deposited on the lake bottom on top of fine-grained sediments. A gravel peak can also be observed in Lake Fryxell core Lz1021 at ca. 115 cm depth (Figure 18), what supports that Lake Washburn might have lost its ice cover for a short period or that the lake level dropped very remarkably at the end of this stage.

4.5.3 Unit III (~11,000-9000 years BP) – Lake Washburn desiccation

The transition from unit II to III at ~11,000 years BP is marked by a change in the lithology of the sediments that indicates a shift in the depositional environment. The abrupt decrease in TIC values and the onset of a sandy accumulation at the end of unit II (Figure 18) indicate an input of freshwater, probably coupled with a slight lake level rise. A warming trend towards the beginning of the Holocene (Steig et al., 2000) may have caused the melting of the RIS and resulted in an enhanced supply of meltwater to the lake. But within this process, the RIS retreat caused an expansion of the lake's surface, while the lake level should be lowered. Hence, elevations of the paleodeltas in eastern Taylor Valley only show a slight increase around 11,500 ^{14}C years BP, followed by a decrease in elevation afterwards (Hall and Denton, 2000b).

The occurrence of volcanic glass in unit III (Figure 18) indicates a still persistent influence of the RIS in Taylor Valley and the existence of Lake Washburn. But due to the fact that the falling lake level of Lake Washburn isolated the Fryxell basin, volcanic material could also derive from the slopes by erosion of older lake sediments (Figure 17).

The transition to a dominantly gravelly sedimentation at ca. 140 cm in unit III of the Lake Hoare record (Figure 14) can be explained by a drop of the lake level that lead to a transition of the coring site to a marginal position in the lake. The sediment sequence of sand coarsening upward into poorly sorted debris is typical for deposits at the margin of proglacial lakes (Hall et al., 2006a). The coring site was located in a zone of high sedimentation. Hall and Denton (2000b) ascribe an intensified deposition to an increased lake ice conveyor activity. The appearance of a second lake-ice conveyor deposits peak at ~11,000 ^{14}C years BP (Hall and Denton, 2000b) seems to fit in this context. But sediment deposition in the marginal parts of Lake Washburn should be high anyway, since the moat was the preferred zone for accumulation of material originating from the ice cover or from the slopes.

Bigger clasts and the absence of fine sediments, which probably have been blown out, are characteristic for lag deposits (Figure 17). Further lake level lowering may be responsible for setting the Lake Hoare coring site subaerially. In this context, it should be mentioned, that ground penetrating radar investigations showed that today Lake Hoare would drain nearly completely into Lake Fryxell by a retreated position of Canada glacier (Burkemper, 2007), which is illustrated in Figure 19.

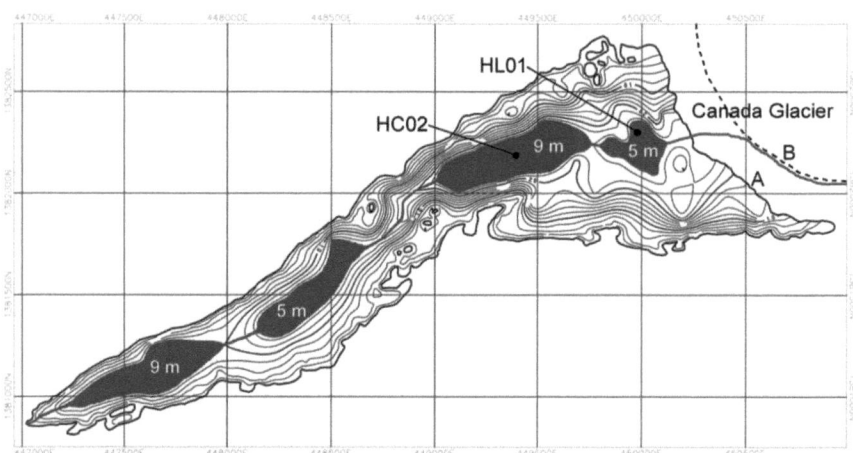

Figure 19. Lake Hoare bathymetry map showing current (solid line at 'A') and retreated (dashed line at 'B') glacier positions, along with remnant ponds (with approximate depths) and streams (Burkemper, 2007). 'HC02' represents the Lz1020 coring location, and 'HL01' represents the location of the annual limnological sampling hut.

For setting the site of core Lz1020 subaerial, lake level must have been lowered below 40 m a.s.l., considering Lake Hoare is located at 73 m a.s.l., and the coring site of Lz1020 at ca. 33 m water depth. Following the occurrence of paleodelta deposits in eastern Taylor Valley, the lake level fell below the critical height of 40 m a.s.l. some time between ~10,500 and 9200 ^{14}C years BP (Hall and Denton, 2000b).

Additionally, ongoing lake level lowering due to evaporation is known from the Fryxell record with indications for an increasing salinity and carbonate precipitation (Wagner et al., 2006). Aragonite horizons with elevated TIC contents were observed in Lake Fryxell sediments, suggesting a fast evaporation of Lake Washburn, firstly dated by Lawrence and Hendy (1985) to ~10,000 years BP and also detected in core Lz1021 by Wagner et al. (2006) around 11,000 and 10,000 ^{14}C years BP (corresponding to ca. 13,000 and 12,000 cal. years BP) (Figure 18). Other studies conducted in the dry valleys region show also indications for an evaporational lowering of the former large proglacial lakes in Miers Valley (Clayton-Greene et al., 1988), Wright Valley (Hall et al., 2001), and Victoria Valley (Hall et al., 2002) between 14,000 and 9,000 ^{14}C years BP.

A small TIC peak in the Lake Hoare record at 140 cm depth supports the assumption of an evaporational lake level lowering (Figure 18). The carbonate precipitation would have stopped, when the coring site was set subaerial, whereas the carbonate precipitation may have continued in the central parts of the lake, as preserved in the Lake Fryxell records. The lower parts of eastern Taylor Valley were still occupied by a lake, which was characterized by a non-perennial ice cover (Figure 17), as indicated by gravels in the sediments of core Lz1021 from Lake Fryxell (Figure 18).

Unit III represents a period, when the RIS was retreating during the warming period at the beginning of the Holocene. The enlarged lake surface area leading to a larger ablation surface promoted an evaporational lowering of the lake level, since the water input of the melting RIS cannot compensate the water loss by evaporation. Additionally, by the retreat of RIS glacial front beyond the Explorer's Cove threshold, the direct meltwater input was reduced and the water level of the thereby isolated lake in the Fryxell basin could not rise above 78 m (corresponding to the minimum elevation of the Explorer's Cove threshold). This would explain, why an enhanced evaporation of Lake Washburn may have occurred during a time, when the climate was getting warmer and likely more humid (Steig et al., 2000).

4.5.4 Unit IV (~9000 years BP to present) – Holocene lake history

Above the significant gravel unit (III), sandy material dominates the sediment deposition in unit IV. Sporadic occurrence of gravel indicates low lake levels or a loss of the perennial ice cover (Figure 14). The deposition of unit IV was less influenced by the dynamics of the RIS. This assumption is supported by the gradual disappearance of volcanic glass and the lower contents of amphiboles in the sediments (Figure 14; Figure 18). Since the lake level fell below the 78 m high threshold between the Fryxell and Explorer's Cove basins (Figure 17), the lake would have lost its direct connection to the ice sheet. In this context, recent studies on marine sediments in McMurdo Sound showed that the RIS grounding line already retreated south of Ross Island between 11,000 and 10,000 ^{14}C years BP and that the ice calving line was pinned to the north of Ross Island at about 9000 ^{14}C years BP (McKay et al., 2008). This would mean that the flowline bringing volcanic material to Taylor Valley broke off around this time. However, geomorphological studies (Hall and Denton, 2000b; Hall et al., 2000) document that a grounded ice sheet at the mouth of Taylor Valley was existing at least until 8340 ^{14}C years BP, and the delivery of volcanic glass to eastern Taylor Valley may have lasted at least until 8700 ^{14}C years BP. In Lake Hoare core Lz1020, volcanic glass is detectable until ca. 60 cm depth, corresponding to an age of maximal ~5500 years BP (Figure 18), a period when RIS have been definitely retreated (cf. Hall and Denton, 2000b). Therefore volcanic glass in the lake sediments may also be reworked Lake Washburn sediments from the valley slopes.

Around 6000 years BP (~60 cm depth), increasing feldspar amounts in Lake Hoare record (Figure 14) may represent a change in the source area, related to an advance of Canada Glacier, bringing material from Granite Harbor Complex to Lake Hoare. This hypothesis is supported by the decoupled preservation of TIC in sediments of lakes Hoare and Fryxell after 6000 years BP (Figure 18), and is in accordance with the advance of alpine glaciers in Taylor Valley (Alpine I drift) during the Mid-Holocene (Denton et al., 1989). The final retreat of the RIS from Taylor Valley (Hall and Denton, 2000b) and the establishment of open marine conditions in the Ross Sea (Hall et al., 2006b; McKay et al., 2008), supported by an mid-Holocene temperature rise, may have lead to moister conditions in the dry valleys contributing to an advance of the alpine glaciers (Steig et al., 2000). The advance of Canada Glacier could have led to a separation of lakes Fryxell and Hoare during the middle Holocene (Figure 17). Since no indications for draining events can be observed in the sediment records afterwards (Figure 18), it can be assumed, that from this time, Canada Glacier did not retreated beyond its present known position ensuring a damming of Lake Hoare (see also Figure 19).

The last significant TIC peak in the Lake Hoare record coincides with a TIC peak in the Lake Fryxell record between 6000 and 4000 cal. years BP (Figure 18). This can probably be traced back to a further evaporational lake level lowering and even to a desiccation event. However, in the sedimentological record, no direct indications, e.g. an erosive hiatus, can be found for such an event. Only the lacking TIC in the sediments of the upper 40 cm of the core and the decrease in the CTDS values (Figure 14) argue for a refilling of the basin with freshwater. Lyons et al. (1998) assumed that Lake Hoare completely desiccated around 2000-1200 years ago based on molecular diffusion models. Even though this event postdates our interpretation, it is most likely that the Lake Hoare basin nearly desiccated for at least once during Holocene times. For Lake Vanda, which is located in the close-by Wright Valley, Lyons et al. (1985) reconstructed a lake level lowering due to evaporation at ~4000 years BP. They suggest that colder and drier climate conditions may have lead to a reduced meltwater supply to the lakes. The significant carbonate precipitation in the Fryxell core Lz1020 (Figure 18) around 5000 cal. years BP (Wagner et al., 2006) could also be traced back to a period of evaporation.

The Lake Hoare sequence shows no significant changes in the upper part of unit IV. On the other hand, the Lake Fryxell record indicates significant environmental changes (Figure 18) that are either not preserved in the Lake Hoare record or have not influenced the sedimentation in the lake.

In regard to bottom water conditions, a change in the TS contents of core Lz1020 can be observed. An increase from ca. 20 cm depth (Figure 14) probably indicates anoxic conditions in the bottom waters of Lake Hoare or at least below the sediment/water interface, which may have established as a result of the lake level rising due to an increased meltwater supply. The latter could have been coupled with an increased fluvial activity. Streams possibly transported reworked material from Lake Washburn deposits on the valley slopes into the lake, indicated in core Lz1020 by the occurrence of volcanic glass and peaks in the amphibole and feldspar contents during this period (Figure 14). A higher lake level and lacking mixing in the water column, resulting from the development of a perennial ice cover, may have lead to anoxic conditions in the bottom waters, supporting the formation of pyrite. In this context, the TS peak at ca. 15 cm depth (Figure 14) could be a sign for a highstand of Lake Hoare. Comparable lake level highstands in the dry valleys, e.g. described for Lake Fryxell and Lake Vanda (Hendy, 2000a), between 3000 and 2000 ^{14}C years BP are ascribed to an elevated meltwater supply. Very warm and humid climate conditions are reconstructed between ca. 2300 and 1100 ^{14}C years BP by Hall et al. (2006b), when the greatest sea ice decline in the

Holocene was responsible for a considerable expansion of seal colonies on the Victoria Land Coast ("Seal Optimum").

This period was followed by a significant lake level lowering, as suggested by Lyons et al. (1998) on the basis of stable isotope data. In combination with an evaporation of Taylor Valley lakes to small hypersaline ponds, Lake Hoare is supposed to have completely dried out about 1200 years ago (Lyons et al., 1998). Although investigations of Spaulding et al. (1997) on Lake Hoare sediments reveal indications for a complete evaporation by changes in the diatom assemblage, we assume that Lake Hoare wasn't desiccated at least in the last 2500 years (Figure 17). Based on the high-resolution radiocarbon dating on microbial mats conducted by Doran et al. (1999), a relatively continuous sedimentation is indicated at the sampling site DH 2 in Lake Hoare (Figure 10a), current at ~11 m water depth, since 2500 ^{14}C years BP. Furthermore, evaporation processes and related increased salinity are not observable in core Lz1020 from Lake Hoare, unlike in the Lake Fryxell record, where a TIC peak and a small gravel peak occur in core Lz1021 (Figure 18) around 2000 cal. years BP (Wagner et al., 2006). Only the abrupt decrease in TS contents at ca. 10 cm in core Lz1020 (Figure 18) could give indications for a small water level change in Lake Hoare.

With the subsequent refilling of the lake, recent conditions established in the lake. Growth of algae mats due to an increased bioproductivity and the resulting preservation of organic matter led to increased TOC, TN and TS contents in the surface sediments (Figure 14). Increasing TS contents in the uppermost 10 cm of core Lz1020 (Figure 18) would also imply that the separation of oxic surface waters and anoxic bottom waters by the establishment of an oxycline have taken place in the recent past. The sandy material in the surface sediments, which is interstratified with microbial mats (Figure 14), reflects the episodic character of clastic deposition in Lake Hoare today, mainly controlled by the perennial ice cover (Squyres et al., 1991).

4.6 Conclusions

Detailed investigation of up to 2.3 m long sediment sequences (Lz1020) recovered from Lake Hoare, Taylor Valley, southern Victoria Land, Antarctica, using chronological, sedimentological, and biogeochemical methods, provides new information about the environmental history of the region. Changes in the reservoir effect have likely occurred with regard to lake level oscillations in the past and therewith connected changes in the glacial melt supply and hydrology of the lake. For these reasons, the differentiated units representing different stages in lake history can only be approximately chronologically constrained.

Nevertheless, the sediment sequence Lz1020 provides a record, which penetrates back to the late Weichselian, to ~17,000 years BP. The data indicate that the proglacial Lake Washburn had occupied the lower part of Taylor Valley during the end of the last glaciation. This lake was dammed by an advanced ice sheet blocking the mouth of the valley and underwent significant lake level oscillations. Lake level fluctuations in the late Weichselian were most likely related to climatic changes, and variations in the extent of the grounded Ross Ice Sheet (RIS). The final evaporative lowering of Lake Washburn is indicated by our data to have started ~14,000 years BP. The Lake Hoare record suggests a very low water level of Lake Washburn some time around 9000 years BP. In comparison to the Lake Fryxell record, our data also show the relevance of location in a proglacial lake and of the environment of the lake resulting in slightly differing sedimentological records.

At the beginning of the Holocene, a significant change in the environment may have occurred in eastern Taylor Valley. The RIS was in retreat and so the lake history became decoupled from its dynamics, whereas the role of the alpine glaciers was getting more important during the Holocene. The lacustrine character of unit IV suggests that a lake existed in the Hoare basin during the Holocene, as a remnant of Lake Washburn. Environmental conditions comparable to those of today likely established with an advance of Canada Glacier around 6000 years BP, whereby Lake Hoare was dammed and therefore separated from Lake Fryxell.

The sediment sequence from Lake Hoare is an useful archive for the reconstruction of the regional paleoenvironment and paleoclimate by providing crucial information, but has to be interpreted in context with those from other archives, like ice core, marine, and terrestrial records.

5 The East Lake Bonney record

5.1 Introduction

Salt lakes are common features in hyperarid, warm climates, but can also be found in extremely cold regions, like the McMurdo Dry Valleys in southern Victoria Land, Antarctica (Matsubaya et al., 1979). Lake Bonney, located in Taylor Valley, is characterized by hypersaline bottom waters and thick salt deposits in its eastern lobe. The origin of the evaporites in Lake Bonney are subject of discussion (e.g., Hendy et al., 1977; Matsubaya et al., 1979; Keys and Williams, 1981; Poreda et al., 2004), but agree that the salts are a legacy of the lake's past (Neumann et al., 2004).

Since the lake occupies the upper valley and lies adjacent to the Taylor Glacier, the late Quaternary history of the Bonney basin is closely linked to the glacial history of Taylor Valley. During the penultimate interglaciation (MIS 5), Taylor Glacier advanced far eastward into Taylor Valley. As indicated by glacial deposits called Bonney drift, the glacier overrode and likely reshaped the Bonney basin (Higgins et al., 2000b). During the last glacial period, Taylor Glacier was in a retreated position and a large lake proglacial to an advanced Ross Sea Ice Sheet (RIS) in the McMurdo Sound occupied Taylor Valley (Hall et al., 2000). Lake Washburn had a maximum lake level of up to 300 m, suggested by dated paleodeltas found on the valley slopes, and occupied the Bonney basin at least until ca. 12,000 ^{14}C years BP, corresponding to the age of the youngest described delta (Hall and Denton, 2000b). Enhanced evaporation led to a lake level drop in the earlier Holocene. Based on molecular and isotopic diffusion rates in the water column, lake level rising of the western lobe finally lead to flooding of the east lobe from ca. 3000 years ago (Poreda et al., 2004). The establishment of a freshwater lens above the saline waters was responsible for the formation of a perennial ice cover ~200 years ago (Poreda et al., 2004).

Former investigations on Lake Bonney were concentrated on its current biogeochemical and physical properties, mainly within the scope of the McMurdo Longterm Ecological Research (LTER) Program (e.g., Priscu, 1998). Wilson et al. (1974) firstly recovered several short cores from the east lobe of Lake Bonney, which revealed the existence of a salt crust mainly consisting of halite. Further investigations on this material focused on the mineralogy of the evaporites (e.g., Craig et al., 1974), and were used to date geochemical events in Lake Bonney (Hendy et al., 1977; Hendy, 2000a).

During the expedition in austral summer 2002/2003, a 2.7 m long core (Lz1023) was recovered from the eastern lobe of Lake Bonney (Wagner, 2003). It is the longest sediment

5 The East Lake Bonney record

sequence so far recovered from the salt crust in East Lake Bonney. This chapter provides a detailed sedimentological characterization of core Lz1023 and first paleoenvironmental implications of the East Lake Bonney record.

5.2 Lake Bonney

Lake Bonney is located at 57 m a.s.l in upper Taylor Valley, adjacent to Taylor Glacier (Figure 20; Table 2). The lake is separated in a western (WLB) and an eastern lobe (ELB) by a subaquatic threshold at ca. 12 m water depth, which is an extension of the Bonney Riegel (Figure 20c). Both lobes have a ca. 900 m width. The west lobe with 2.6 km length is nearly half as long as the east lobe (4.8 km in length). The bathymetry of the lake shows steep slopes and deep basins in both lobes. Maximum water depths of ~40 m are located in the center of ELB and in the western part of WLB adjacent to the tongue of Taylor Glacier (Figure 21a). The lake level has been rising during the last century (Chinn, 1993). The perennial ice cover of both lakes is smooth and has a thickness of 3.0-4.5 m on average (Table 2). Water temperature below the ice cover is ~0°C. In East Lake Bonney it rises up to a maximum of nearly 5°C at 18 m depth (Figure 21b). Subsequently, water temperature gradually decreases and reaches -1°C at the lake bottom (Lyons and Welch, 2007).

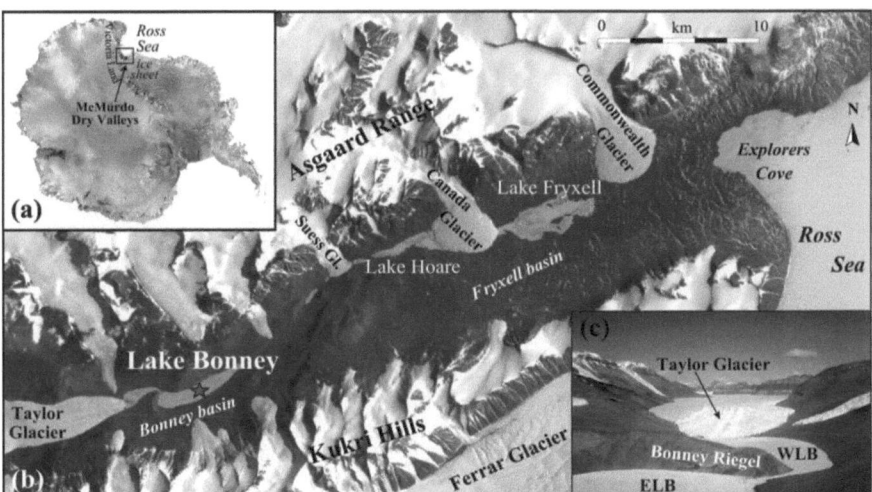

Figure 20. Study site: (a) Location of McMurdo Dry Valleys in Antarctica (U.S. Geological Survey, 2006). (b) Landsat 7 satellite image of Taylor Valley, southern Victoria Land, Antarctica, indicating the location of the most important lakes and glaciers (U.S. Geological Survey, 2007). The red star in East Lake Bonney indicates the coring location Lz1023. (c) Photograph of Lake Bonney and Taylor Glacier (source: Gunn, 2006b).

5 The East Lake Bonney record

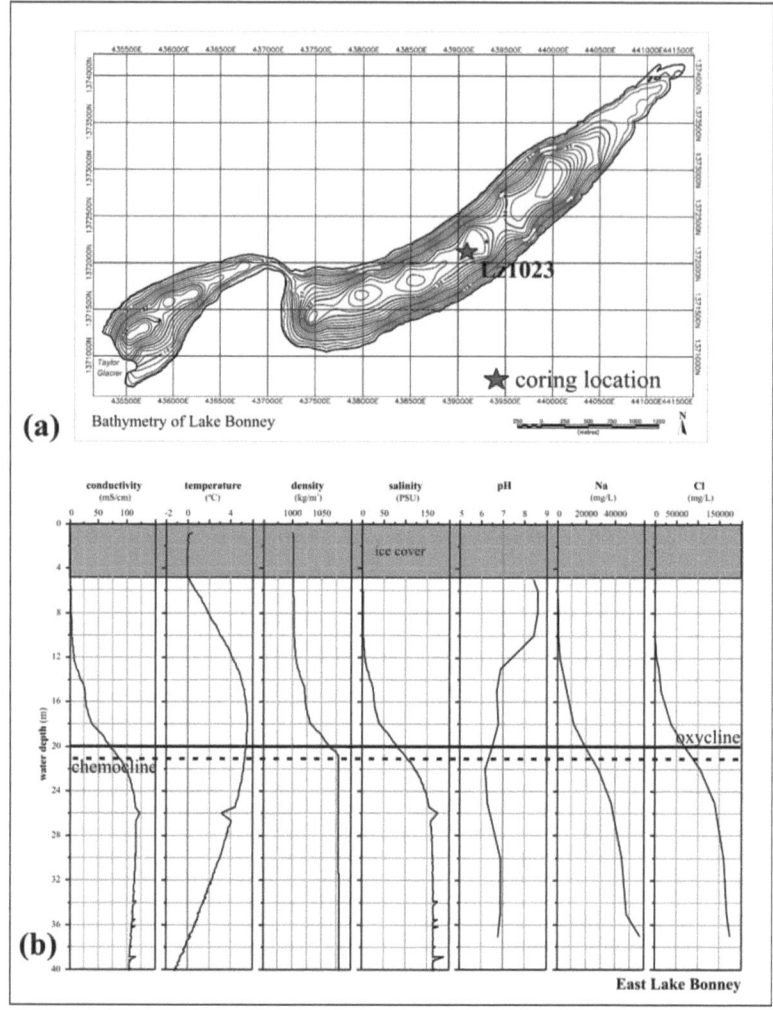

Figure 21. (a) Lake Bonney bathymetry (Schmok and Waddington, 1996) and (b) waterprofile of East Lake Bonney (McMurdo LTER data, season 2002/2003; Lyons and Welch, 2007).

Water input mainly originates from meltwater of Taylor Glacier, but also from streams that deliver meltwaters from adjacent alpine glaciers (e.g., Matterhorn, Sollas, Hughes or Rhone Glacier) (Matsubaya et al., 1979). Special feature of Lake Bonney is the hypersaline bottom water with values of >150 psu (Lyons and Welch, 2007). Recent studies of Mikucki et al. (2004) reveal the Blood Falls as possible source, a discharge from a saline iron rich water body, being expected of lying beneath Taylor Glacier (Hubbard et al., 2004). More recent studies of Lyons et al. (2005) suggest that the saline waters in Lake Bonney originate from ancient seawater, which was modified over time by addition of salts from atmospheric precipitation as well as from leaching soils, by freeze drying, and by subsequent loss of solutes via mineral precipitation, especially in the case of East Lake Bonney.

In addition, Lake Bonney is characterized by an oversaturation with respect to sodium chloride in its bottom waters (Hendy et al., 1977). Salinity as well as conductivity and density are decreasing upwards in the water column (Figure 21b), indicating diffusion processes coming from the hypersaline bottom waters. A chemocline is established at ca. 22 m water depth in East Lake Bonney (Lawson et al., 2004). Below the ice cover, lake water is characterized by low conductivity (<4 mS/cm) and salinity (<4 psu) indicating a freshwater lens at the surface of the lake (Figure 21b). Oxygenated waters can only be found in the upper 20 m of the water column (oxycline) in both lobes and originate from meltwater input from the catchment (Lawson et al., 2004).

5.3 Material and methods

From East Lake Bonney, the 2.7 m long piston core Lz1023 (location: 77°42.892 S, 162°26.216 E) was recovered from 38.3 m water depth (Wagner, 2003). At the University of Illinois, Chicago, in frozen state, the core was divided lengthwise into two halves, then described and photographically documented. Further on, one half of the core was split into two quarters for subsampling, which was carried out in 2 cm intervals, while the core was thawing. Analyses of the cores were conducted according to Figure 22. The water content was determined by the weight loss during freeze-drying of the subsamples.

For analyzing the ionic composition of the salts, the dried sediment samples were diluted in deionized water and then filtrated through a 0.45 um filter for separating the clastic material. The residual liquid, which contains the dissolved salts originating from the crystals and the formerly dried pore waters, was analyzed with respect to major cations and anions. At the University of Jena, sodium (Na) and potassium (K) were measured by atomic adsorption spectrometry (AAS) with an AA 6800 (Shimadzu Corp.), and chloride concentrations were determined using an ion chromatograph DX-120 (Dionex Corp.).

In addition, aliquots of the clastic fraction were separated in order to conduct grain-size analyses with a laser particle size analyzer 1180 (CILAS Corp.). Since the particle size analyzes do not measure the gravel fraction accurately, the clastic fraction samples were investigated due to the occurrence of grains with diameters larger than 2 mm.

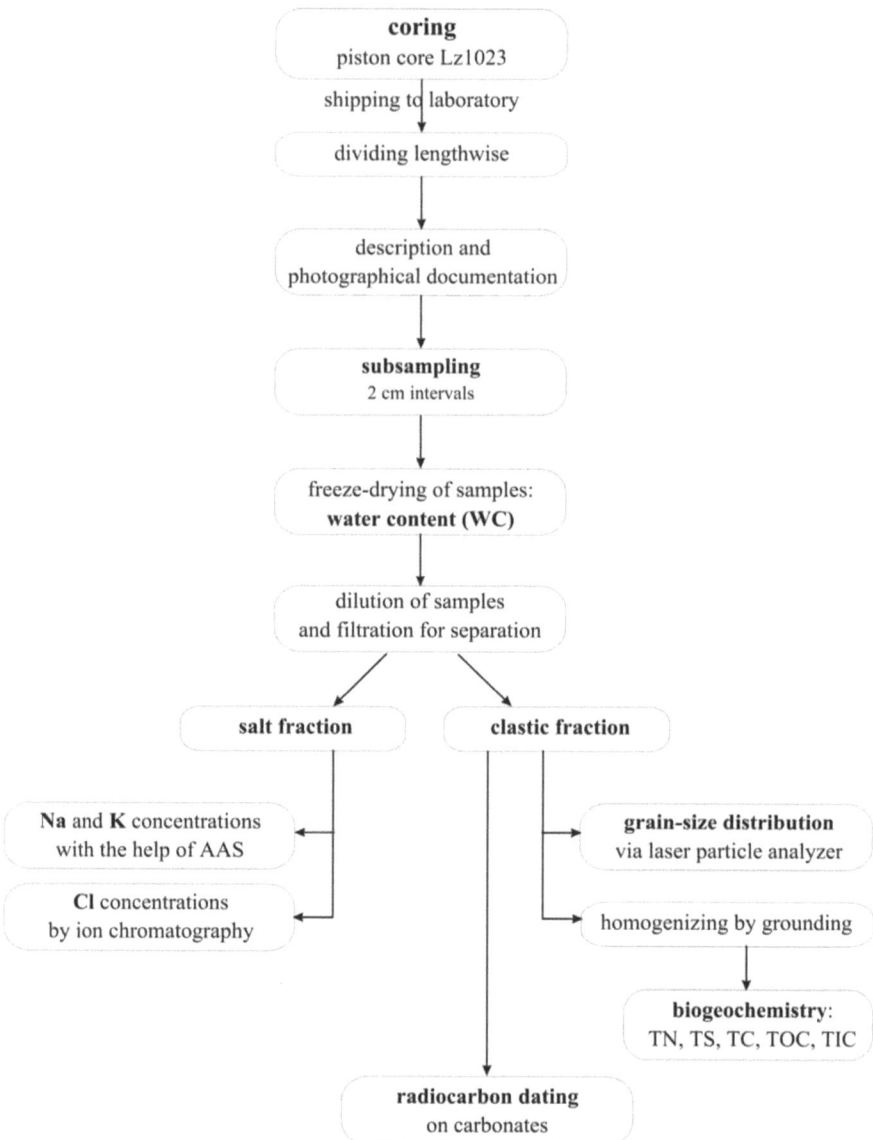

Figure 22. Flowchart of methods for analyzing core Lz1023 from East Lake Bonney.

For biogeochemical analyses, an aliquot of each clastic subsample was ground to <63 µm and homogenized. Total sulfur (TS) and total nitrogen (TN) contents were measured with an Elementar III (VARIO Corp.) analyzer. Since the measured concentrations of TN were very low, partly below the detection limit, TN is not taken into consideration. Total carbon (TC) and total inorganic carbon (TIC) contents were determined with a DIMATOC 100 (DIMATEC Corp.) analyzer. The total organic carbon (TOC) content was calculated by the difference of TC and TIC contents.

Radiocarbon dating was conducted by accelerator mass spectrometry (AMS) at the Leibniz Laboratory for Radiometric Dating and Isotope Research in Kiel, Germany. Since the amounts of the clastic fraction and TOC contents are very low, a sample with a sufficient amount of TIC was selected from 170-174 cm depth for dating the carbonates (Table 5). The repeated dating on the sample yielded ages of 10940 ± 100 and 10830 ± 60 ^{14}C years BP, respectively.

Table 5. Radiocarbon dates of carbonates from a bulk sediment sample of core Lz1023, East Lake Bonney.

Sample	Core	Depth (cm)	Material	C (mg)	δ^{13}C (‰)	^{14}C age (^{14}C years BP)
KIA 37156	Lz1023	170-174	carbonate	0.2	9.21	10940 ± 100
			carbonate	0.9	6.01	10830 ± 60

5.4 Results and discussion

5.4.1 Stratigraphy

The sediments of East Lake Bonney (ELB) are composed of three major components: water, salts, which predominately consist of idiomorphic halite (NaCl, Figure 23), and the clastic fraction (Figure 24). The salt crystals are varying in their sizes from very small (<1 mm) up to 2 cm edge length. The water content of core Lz1023 averages at ca. 10% in the salt crystal dominated parts. Above 60 cm depth, the water content gradually increases to ca. 65 % at the top (Figure 24). As clastic fraction, we define the fraction, which is left after dissolving the halite in a sample. The amount of clastic components in ELB is low with less than 10 %. Only between 40 and 20 cm depth, the amount of clastics reaches ca. 25 % (Figure 24). Values of total sulfur (TS) are pretty low, except for the upper part of the core and thin layers in the lower core. The amount of total carbon (TC) is distinctly higher, reaching ~6 % of the clastic fraction at its maximum. The main proportion of carbon (>60 %) is formed by inorganic material (TIC) (Figure 24). Based on its sedimentological properties, the core can be differentiated in four distinct units.

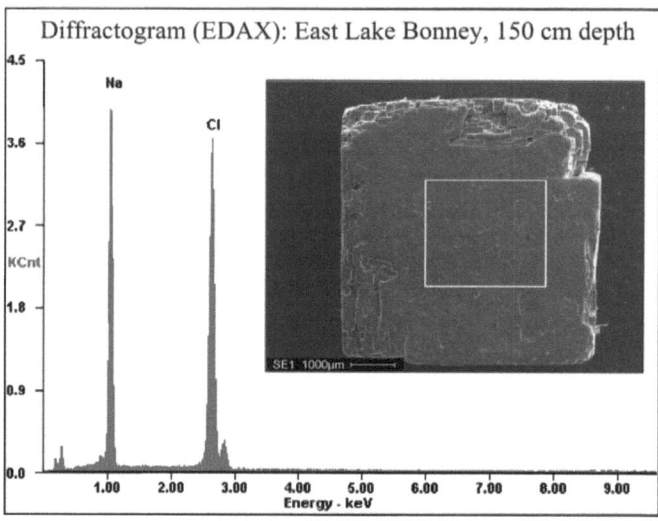

Figure 23. Diffractogram and SEM photo of salt crystal from core Lz1023, 150 cm depth.

5.4.1.1 Unit 1

Unit 1 ranges from 270 cm to 210 cm depth (Figure 24). The core base mainly shows medium-sized salt crystals, and clastic material can sparsely occur. From 260 to 250 cm depth, large crystals dominate the composition of the sediment. On top of this horizon, a mixture of various sized crystals can be observed. The size of the crystals varies between 1 mm and 1 cm (average: ~0.5 cm). Above 220 cm depth, the size of the salt crystals decreases to a dimension, where the crystals can hardly be seen macroscopically. Analyses on the salts show the dominance of sodium (Na) and chloride (Cl) with ca. 35 % and nearly 60 %, respectively. The molar ratio Na/Cl is relatively constant (~0.9). In the salt fraction, potassium (K) plays a minor role with ca. 0.02 % (Figure 24).

The clastic fraction mainly consists of sand. The proportion of silt and clay is slightly decreasing upward. Apart from a single evidence for grains >2 mm occurring in the lower part of unit 1 (250 cm depth), gravels are lacking. Contents of total carbon (TC) vary between 4 and 6 %. The carbon mostly consists of total inorganic carbon (TIC), which averages around 5 %. Sulfur (TS) is lacking (Figure 24).

5 The East Lake Bonney record

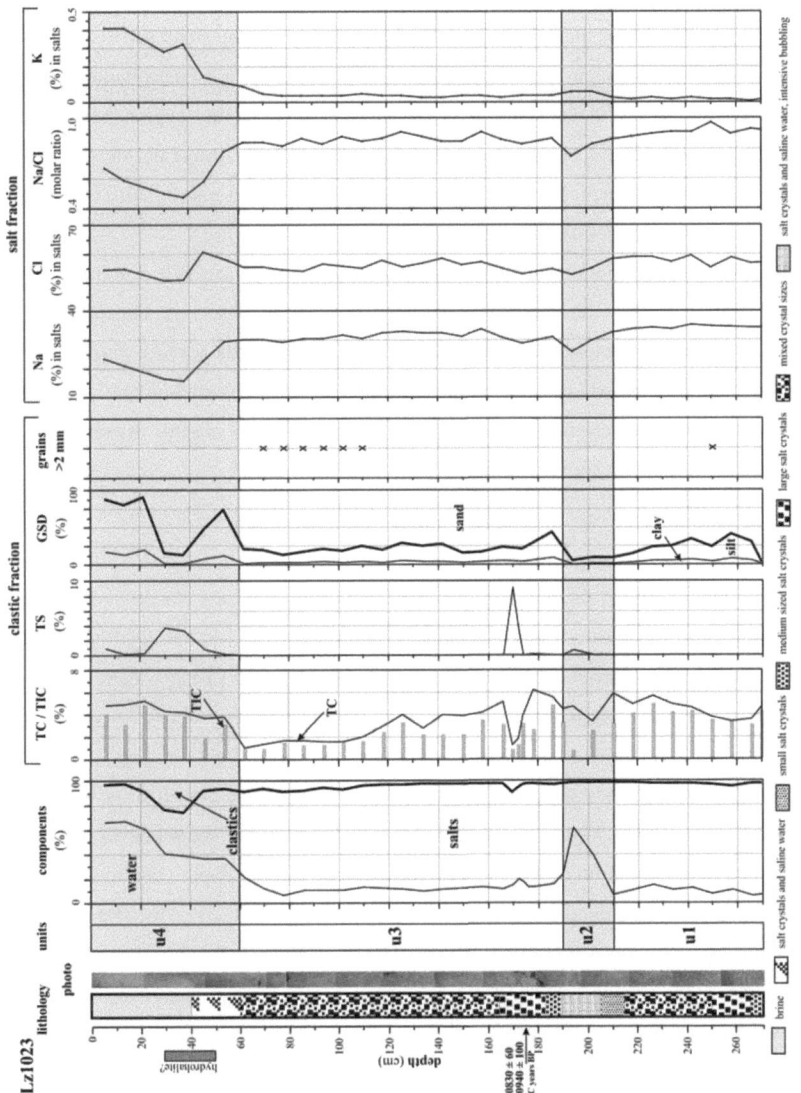

Figure 24. Characterization of core Lz1023: Lithology, components, water content, total organic carbon (TOC), and total inorganic carbon (TIC), total sulfur (TS), grain-size distribution (GSD with clay, silt and sand from left to right) and index for occurrence of grains >2 mm (crosses indicate the presence of grains >2 mm in the investigated samples) in the clastic fraction, sodium (Na) and chloride (Cl) concentrations, molar ratio of sodium to chloride (Na/Cl) and potassium (K) concentrations in the salt fraction.

5.4.1.2 Unit 2

Unit 2 (210 to 190 cm depth) represents an almost homogenous layer (Figure 24). During thawing of the core, this layer showed a remarkable bubbling and a very high release of water. The high water content was responsible for a partial dissolution of very small salt crystals.

Maxima of the water content between 210 and 190 cm depth can be explained by the release of water during the decomposition of the gashydrates. Another explanation for the elevated water contents could be the decomposition of metastable minerals, like ikaite or hydrohalite, but would not explain the intense bubbling. However, in unit 2, there is no clear evidence for the kind of (mineral) phase.

Regarding to the salt fraction, sodium (Na) and chloride (Cl) are still dominating the composition, but a decrease in the Na/Cl molar ratio can be observed, concurrently K concentrations increase in this section (Figure 24). Clastic components are rare in unit 2. The sparse material mainly consists of sand, gravels are lacking. The carbon content is variable between 3 and 6 % of the clastic fraction. In one sample (194 cm depth), organic carbon seems to dominate and sulfur shows a small peak (Figure 24).

5.4.1.3 Unit 3

Unit 3 (190-60 cm depth) predominately consists of salts, in which clastic particles can occur. The amount of the clastic fraction is slightly increasing from 120 cm depth to the top. The water content (WC) is around 10 %. A small peak of WC at ca. 170 cm depth coincides with a minor spike in the clastic fraction. The WC slightly increases above 80 cm depth, (Figure 24).

The salt fraction of unit 3 is similar to unit 1 in its variation of the crystal sizes, and in its constant Na/Cl composition. Medium-sized salt crystals dominate the sediments in the lower part. Larger salt crystals can be observed between 182 and 165 cm depth. The main part of unit 3, from 165 to 60 cm depth, consists of mixed sized salt crystals. Amounts of K are relatively constant with values between 0.03 and 0.04 % of the salt fraction throughout this unit, but increase at the top to nearly 0.1 % (Figure 24).

The clastic fraction mainly consists of sand. The proportion of clay and silt is around 25 %, except for higher amounts in the lower part of unit 3. Grains with diameters larger than 2 mm occur in the upper part of unit 3 between 110 and 70 cm depth (Figure 24). Amounts of carbon are variable in unit 3. In general, total carbon (TC) content is decreasing from ca. 7 % to 1 %, with relatively constant values of 1-2 % between 120 and 60 cm depth. With few exceptions, the main part consists of inorganic carbon (TIC). A significant drop of the carbon values to less than 2 % can be observed at ca. 170 cm depth, whereas sulfur (TS) contents show maximum values of about 9 % at this depth. The remaining part of unit 3 shows very low TS contents (Figure 24).

5.4.1.4 Unit 4

Unit 4 reaches from 60 cm depth to the top of core Lz1023 (Figure 24). At ca. 60 cm depth, the composition changes and macroscopic salt crystals are occurring only sparsely. The salt content is high anyway, since very small salt crystals could be observed in the frozen core. The brownish color of the core between 60 and 20 cm depth can be traced back to an increased amount of clastic material in this section (Figure 24). The upper 20 cm of core Lz1023 consist of a white homogenous paste with less clastic material, interpreted as frozen brine.

The increasing proportion of water towards the top of the core, where the values reach a maximum of ca. 65 % (Figure 24), can be traced back to a gradual transition from sediment to the lake's bottom water, which is supersaturated with salts. The weight percentage of salts is remarkably decreasing in the uppermost part of the core, which supports the assumption that unit 4 represents the transition between sediment and the overlying brine. A decrease in Na concentrations can be observed, especially between 40 and 30 cm depth, where Na concentrations drop to ca. 15 % of the salts. Potassium (K) concentrations increase and show maximum values of up to ca. 0.4 % of the salt fraction (Figure 24).

The clastic fraction in unit 4 is characterized by high percentages of silty material. Between 60 and 40 cm depth, silt and clay particles amount up to 80 % of the clastics. Sandy material dominates between 40 and 30 cm depth. The uppermost part of the core is characterized by the dominance of silt (Figure 24). The relatively high amounts of silt-sized clastics could be related to an accumulation of the lakes suspension material in the high-density brine. Carbon is present, mainly as inorganic carbon, and amounts 4-5 % of the clastic fraction. Contents of sulfur (TS) show relatively high values, with an increase to nearly 4 % between 40 and 30 cm depth (Figure 24).

5.4.2 Clastic fraction

Except for the uppermost part of the core, the percentage of the clastic fraction is very small. The constant low incorporation of clastic material in the salts could reflect relatively continuous precipitation conditions. Desiccation events would result in interruption of the salt crust by mud layers (Lowenstein and Hardie, 1985).

The high clastic amounts in unit 4 (Figure 24) could indicate that either dissolution processes have taken place, whereby residual clastic material has been accumulated, or continuous precipitation of salts has been interrupted. In case of the latter reason, suspended material could settle through the water column and could deposit at the lake bottom above the actual salt crust. However, both explanations can be associated with a freshwater input to the lake.

Grain size analysis on the clastic fraction has shown that it mainly consists of sand size. Aeolian transport is the dominant mode in Taylor Valley. The high wind velocities (Doran et al., 2002a) can move grains up to sand size. Since the ice cover of Lake Bonney is relatively smooth, only small amounts of the wind-transported material can be trapped in and on the ice cover. By freezing and thawing processes, this material can pass through the ice, whereby Hendy (2000b) suggest that grains with diameters larger than 1 cm cannot melt through the perennial ice covers of today dry valley lakes. Meltwater inflow via streams is responsible for a supply of fine-grained suspension load, which can be regarded as main source for silt and clay in the clastic fraction of core Lz1023. Therefore, higher amounts of silt and clay in the clastic fraction, e.g. occurring in unit 4 (Figure 24), point to an elevated meltwater supply to the lake. The occurrence of gravels in the sediment could indicate an episodic ice cover on East Lake Bonney. Coarse-grained material could have been trapped on or in floating ice and could have dropped into the lake during summer, when the ice melted. Since the section between 110 and 70 cm depth in unit 3 shows continuous, even though sparse occurrence of grains with diameters >2 mm (Figure 24), the correlated deposition period of these sediments could have been characterized by an establishment of a non-perennial ice cover. However, it is assumed that an perennial ice cover was formed on East Lake Bonney not until 200 years ago (Poreda et al., 2004).

The clastic fraction show elevated amounts in TC and sporadic in TS. However, the general low contents of these parameters are common in Antarctic lake sediments (Hodgson et al., 2004). In East Lake Bonney, organic material is lacking or negligible, since TN was hardly detectable and TOC is generally less than 1 % (Figure 24). This is due to the very low bioproductivity in Lake Bonney and the decomposition of organic material after sedimentation (Lawson et al., 2004). The main part of TC is formed by total inorganic carbon (TIC). Its contents in the clastic fraction are high and variable. TIC occurs most likely in form of calcite or aragonite. Sediments of other dry valley lakes, e.g. Lake Fryxell sediments, show carbonates associated with evaporation events (Lawrence and Hendy, 1989; Wagner et al., 2006). The difference of East Lake Bonney in comparison to the other Taylor Valley lakes is its hypersaline monimolimnion. Today, the bottom waters of East Lake Bonney are supersaturated with respect to calcite (Neumann et al., 2004), and precipitation of calcium carbonates is very likely. For these reasons, increased TIC contents may rather reflect changes in the composition of the bottom water brine than evaporation events. The elevated amounts of sulfur in the sediments may be related to the occurrence of gypsum. A correlation with organic matter is unlikely (Figure 25), but cannot be excluded. Craig et al. (1974) found evidence for gypsum and also aragonite in East Lake Bonney at the sediment surface. The precipitation of calcium carbonates and gypsum is also very likely under current conditions of the brine (see discussion below).

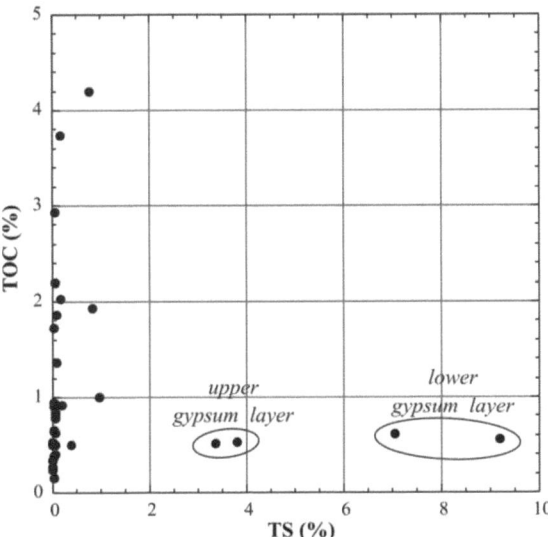

Figure 25. Total organic carbon versus total sulfur values from East Lake Bonney core Lz1023.

5.4.3 Salt fraction

East Lake Bonney (ELB) can be described as a perennial salt lake, which is hydrologically closed and chemically stratified with a dense brine at the lake bottom overlain by fresher surface water (Eugster, 1980). The origin of the brine is not definitely clarified. The chemical composition of its water indicates a marine origin (Matsubaya et al., 1979). Additional sources for solutes are products of chemical weathering in the catchment (Keys and Williams, 1981). Evaporation processes and cryo-concentration of freshwater have played an important role in the evolution of the hypersaline brine in ELB (Poreda et al., 2004). Due to the dominance of sodium and chloride ions, halite precipitation from the brine is responsible for the formation of the more than 2 m thick salt crust at the bottom of ELB. Analyses show that core Lz1023 mainly consists of halite with sparse occurrence of clastic components. Idiomorphy of the cubic salt crystals (Figure 23) suggest a slow process of crystallization (King, 2005).

A remarkable feature of core Lz1023 is the periodical succession of the crystal sizes in units 1 and 3, which probably indicates a cyclic evaporation. Both units show medium-sized crystals in the lower part, followed by a horizon with large crystals and a thick section consisting of mixed crystal sizes (Figure 24). The size of salt minerals gives indication for conditions of crystal growth. Occurrence of large crystals indicates long-term stable concentration for crystal growth in the brine. Mixed crystal sizes can be traced back to changing brine conditions with contemporaneous nucleation and crystal growth (Lowenstein and Hardie, 1985). Alternating conditions of crystal growth can be caused by a variable supply of solutes and by variations in the strength of evaporation processes (Eugster, 1980). Since these processes are complex and their operation in East Lake Bonney is not understood in detail at the moment, it can only be concluded that units 1 and 3 were formed under similar conditions during periods of enhanced evaporation in the basin. The two successions of cyclic precipitation of halite (units 1 and 3) are interrupted by unit 2, between 210 and 190 cm depth (Figure 24). Since the kind of (mineral) phase occurring in unit 2 is unknown, we cannot give an explanation for its formation.

Unit 4 probably shows the transition between sediment and water column. During coring, a subaquatic camera was used showing the sediment surface as a white hard saltpan. However, in the core, a hard surface of salts is formed at 60 cm depth. Above 60 cm depth, high water contents and fewer amounts of salts could be a relict of the coring process, but it is more likely that unit 4 represents the sediment-water transition. The relevance of brine in the upper part of the core can also be seen in the increasing K concentrations in unit 4, since potassium rather occurs in lake and pore waters, than it is incorporated in the halite crystals.

With regard to the mineralogical composition of unit 4, previous investigations on surface sediments of East Lake Bonney described the formation of hydrohalite ($NaCl*2H_2O$) and halite ($NaCl$) in the hypersaline bottom waters of East Lake Bonney (Craig et al., 1974). Hydrohalite, with a theoretical water content of 38.1 %, seems to be a primary precipitate in Lake Bonney, which melts to halite and saturated saline water solution, when the temperature of the bottom water is rising above 0.1 °C (Craig et al., 1974). The section between 50 and 30 cm depth in unit 4 could be related to a hydrohalite horizon (Figure 24), since the water content of the samples is around 40% (Figure 24). Otherwise, carbonates and gypsum are likely in the surficial sediments of ELB, since TIC and TS show maxima in the uppermost part of core Lz1023 (Figure 24). These minerals were also described by Craig et al. (1974), who found small layers of them being interposed in the salt deposits. Additionally, the current composition of the bottom waters with a density of ~1080 kg/m^3 and a salinity of ~160 psu promotes precipitation of gypsum and calcium carbonates after Usiglio (1849, as quoted in Dronkert, 1985). The present lake level rise due to freshwater input (Chinn, 1993) could be responsible for a reversal of the evaporite depositional cycle (Eugster, 1980). The observation of hydrohalite (Craig et al., 1974) does not exclude an interrupted accumulation of halite, since it is possible that present brine conditions can support an equilibrium between precipitation and dissolution. Halite formation and accumulation in ELB have occurred at higher salinity of the brine (Lyons et al., 1999), probably during times of enhanced evaporation in the basin (Poreda et al., 2004).

5.4.4 Chronological considerations

Radiocarbon dating on a sample from 170-174 cm depth yielded an age of 10830 ± 60 and 10940 ± 100 ^{14}C years BP, respectively. Whether these ages are reliable is questionable, since large reservoir effects are common in Antarctic lakes, because of the input of old carbon derived from glacial melt and the lacking exchange with the atmosphere due to their perennial ice covers (Hendy and Hall, 2006).

Furthermore, dating was conducted on carbonates, which originate from dissolved inorganic carbon (DIC) in the water column. Today, deep waters of ELB are supersaturated with respect to calcite (Neumann et al., 2004). Concentrations, radiocarbon ages and δ^{13}C signatures of DIC in East Lake Bonney show remarkable variations (Table 6; Figure 26). Maximum concentrations and highest radiocarbon ages occur in the middle of the water column (Figure 26) and can be traced back to an inflow from West Lake Bonney, where very high reservoir ages of up to ~25,000 ^{14}C years BP were determined for DIC in the water (Lawson Knoepfle et al., in prep.). Younger ages and lower concentrations of DIC beneath the ice cover of East Lake Bonney (Figure 26) reflect the influence of the freshwater lense and stream input (Lawson Knoepfle et al., in prep.). Analyses of δ^{13}C signature of DIC in ELB waters (Neumann et al., 2004) support both the influence of stream input in the surface waters and the chemocline leakage reflected in light δ^{13}C values in the middle of the water column. The bottom waters of ELB with low DIC concentrations are considered as old evapoconcentrated brine waters (Lawson Knoepfle et al., in prep.). For these reasons, DIC of the bottom waters shows the heaviest $\delta^{13}C$ values (Neumann et al., 2004) and yielded ages ranging from 8000 to 10,000 ^{14}C years BP (Doran et al., 1999; Lawson Knoepfle et al., in prep.). These are close to the age obtained by the radiocarbon dating of our sample (Table 5). This suggests a high reservoir effect in the lake today and would mean a younger age of the sediment. For these reasons, the radiocarbon date KIA 37156 from 170-174 cm depth can only be seen as a maximum age for this part of the section.

Table 6. Radiocarbon data of DIC in East Lake Bonney after Lawson Knoepfle (in prep.).

	Dissolved Inorganic Carbon				
	Locale	depth (m)	^{14}C years BP	F_m	$\delta^{13}C$
ELB	Center	6	3491±45	0.65	2.1
	Mid SP	6	3512±35	0.65	2.0
	Center	16	13670±110	0.18	1.1
	Center	21	11262±59	0.25	-0.7
	Mid SP	37	10376±53	0.27	8.0
	Center	39	8581±80	0.34	6.9

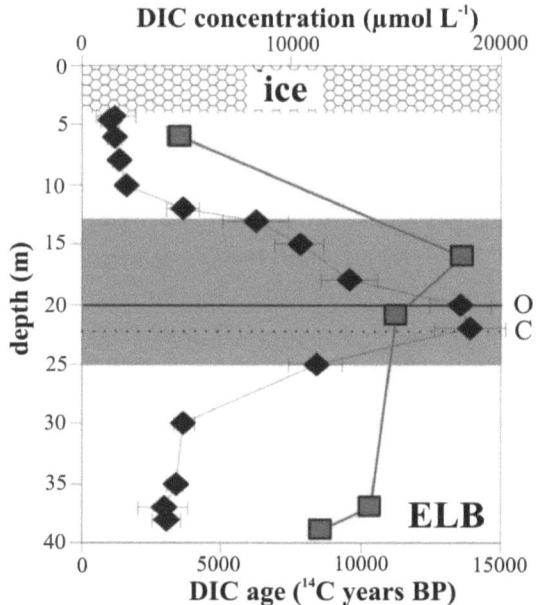

Figure 26. Radiocarbon ages of DIC, and DIC concentrations in East Lake Bonney (ELB) after Lawson Knoepfle (in prep.). The green area marks the $CaCO_3$ unsaturated zone. Black diamonds represent MCM-LTER 10 year average DIC concentrations and red squares are ^{14}C ages. The black line indicates the oxycline (O). The dotted line represents the chemocline (C).

Another attempt for developing a chronology of core Lz1023 is the consideration of accumulation rates in evaporite systems. Very high rates of 2 cm/year for chemical sedimentation were recorded in Freefight Lake in the Great Plains, western Canada (Last, 1993a, b). Considering these very high accumulation rates suggest that the more than 2 m thick salt deposits recovered from East Lake Bonney can be formed within a very short period of time. Although Freefight Lake is similar to East Lake Bonney, such high sedimentation rates are unusual and can only be seen as a maximum value. Schreiber and El Tabakh (2000) suggest lower sedimentation rates of about 1.8 mm/year for stratified hypersaline water bodies, what corresponds to ca. 1500 years for the deposition of 2.7 m salts. Simple calculations based on experiments of Usiglio (1849, as quoted in Dronkert, 1985) and assuming seawater as precipitating liquid and present climate conditions with ablation rates of 150-500 mm/year (Clow et al., 1988) yield 400-1300 years for the deposition of 2.7 m salt crust. However, the history of East Lake Bonney after Poreda et al. (2004) would imply a discontinuous deposition of salts in the past with an alternation of precipitation and dissolution as well as sedimentation breaks. But the relatively constant percentages of clastics throughout units 1, 2 and 3 in core Lz1023 (Figure 24) suggest a continuous sedimentation of halite. Concerning unit 4, age determination can be approximated referring to Hendy (1977; 2000a), who found aragonites and gypsum in the surface sediments in ELB, which yielded ages between 100 and ~2000 years (Table 7).

According to the above, developing a reliable age model for core Lz1023 is very difficult. We can only state that the radiocarbon age KIA 37156 includes a large reservoir effect and that the salt crust was probably formed within few thousand years or less.

Table 7. U/Th ages of former ELB cores (Hendy, 2000a).

East Lake Bonney cores	Material	U/Th age (years)	age after Th correction (years)
core 2, 5-7 cm	aragonite	1800 ± 100	240
core 2, 5-7 cm	gypsum	10,000 ± 550	100
core 2, 11-12 cm	aragonite	1760 ± 130	1130
core 11	aragonite	5600 ± 190	1980

5.5 Paleoenvironmental implications

Since radiocarbon dating on the sediments provided a questionable age, variations in the sedimentological composition can only be used for first paleoenvironmental implications and deducing signs of environmental change. Our investigations on core Lz1023 do not allow direct conclusions about the real age of the recovered salt crust. In the data, only a continuous evaporation phase can be reconstructed, since units 1 to 3 of core Lz1023 (270-60 cm depth) show quite constant amounts of incorporated clastic material (Figure 24) and clastic-rich layers as indicators for desiccation events (c.f. Lowenstein and Hardie, 1985) are lacking.

If we regard the radiocarbon age of sample KIA 37156 as a maximum age, the salt crust was formed at anytime within the last 11,000 years. Evaporation events in western Taylor Valley during the last 11,000 years can be associated either to the final evaporation of Lake Washburn during the Pleistocene-Holocene transition (Hendy, 2000a) or to the late Holocene evaporation event (Lyons et al., 1998). The latter is more reasonable with respect to the expected high reservoir effect contained in date KIA 37156 and the relatively high sedimentation rates in evaporites, which both suggest a younger age of the salt crust.

Another question arises about the origin of the hypersaline waters in East Lake Bonney (ELB), which provide the basis for halite precipitation. Today, saline waters in West Lake Bonney (WLB) derive from the Blood Falls (Mikucki et al., 2004), which could be transferred into the east lobe. Poreda et al. (2004) suggest that cryo-concentration of freshwater inflowing from WLB contributed to the formation of the brine in the east lobe, whereas Lawson-Knoepfle et al. (in prep.) discuss chemocline leakage as source of saline waters in ELB. The marine signature of ELB waters confirms the Blood Falls as major source of the saline waters (Lyons et al., 2005). Only through an overflow from the west lobe, these waters can reach the east lobe.

Due to the climate change after the final retreat of the Ross Sea Ice Sheet from McMurdo Sound between 9400 and 7600 ^{14}C years BP (Conway et al., 1999; Hall and Denton, 2000b), Taylor Glacier and alpine glaciers in the catchment of Lake Bonney advanced (Denton et al., 1989). During this so-called Alpine I drift, Taylor Glacier may have reached its maximum position between 3500 and 2500 ^{14}C years BP (Hall and Denton, 2000b; Higgins et al., 2000a). This was probably the time, when the west lobe was filled up with meltwater from Taylor Glacier to a level (Hendy et al., 1977), that waters spilled over into the east lobe for the first time. Poreda et al. (2004) suggest an overflow from WLB to ELB starting ca. 3000 years ago (Figure 27). Beside the contribution of maximum extent of Taylor Glacier to the lake level rise in WLB (Hendy et al., 1977), evidence from Lake Vanda suggest lake level highstands in the dry valleys region due to more humid climate conditions between 3000 and 2000 years BP (Smith and Friedman, 1993). Following this stage, several archives indicate a climate change to colder and drier conditions between 2000 and 1000 years BP (e.g., Baroni and Orombelli, 1994a; Lyons et al., 1998). The evaporites, occurring in units 1 and 3 of core

Lz1023 (270-60 cm depth), could possibly originate from this dry period, since calculations based on sedimentation rates yield approximate ages of 400-1500 years.

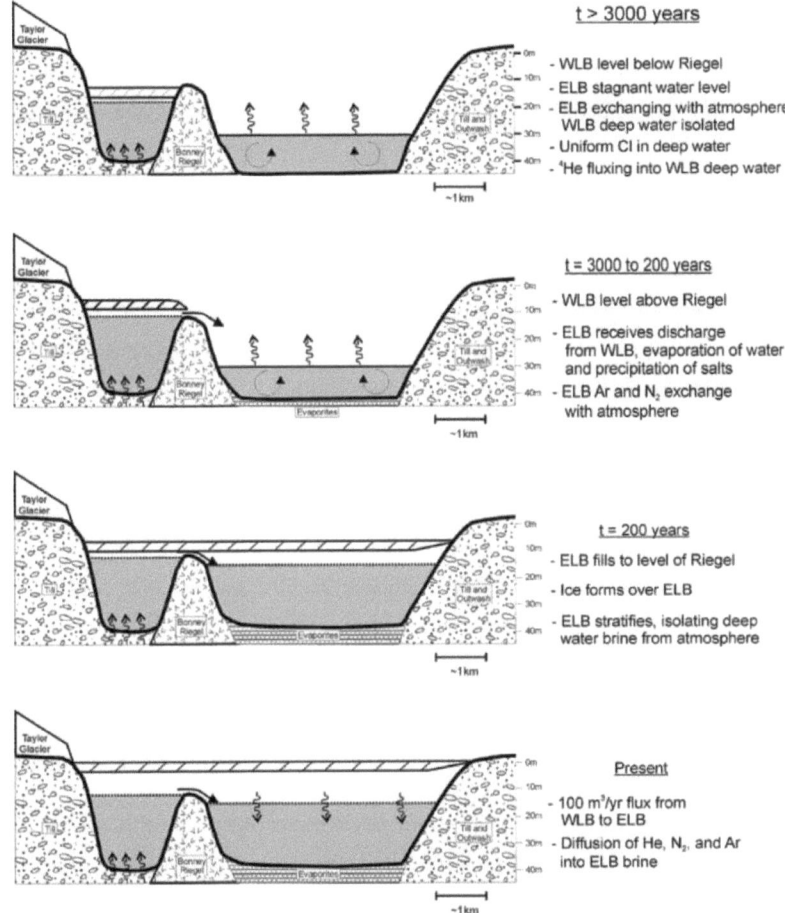

Figure 27. Chronology of events at Lake Bonney during the Holocene (Poreda et al., 2004).

Based on investigations of isotopes in the water column of Lake Bonney, Poreda et al. (2004) suggest an episodic water supply to ELB, and evaporation leading to the formation of salts during the period between 3000 and 200 years ago (Figure 27). Precondition for the precipitation of halite is that either evaporation exceeded the inflow of freshwater or saline waters from the west lobe were spilled into the east lobe for providing a continuous supersaturation of the brine with respect to halite. The variations in crystal sizes observed in unit 1 and 3 of core Lz1023 (Figure 24) could be traced back to variations in the constitution of the brine. Additionally, salt precipitation in cycles and with regard to the occurrence of gypsum and carbonates, also observed by Hendy (1977), indicate variations in the water supply to the lake.

We suppose that units 1, 2, and 3 may originate from some time within the period between 3000 and 200 years ago after Poreda et al. (2004), whereas we do not know the basal age of our core. Furthermore, the conditions for the formation of unit 2 of our core (Lz1023) are not clear at the moment. Regular occurrence of gravels between 110 and 70 cm depth in unit 3 of core Lz1023 (Figure 24) could be traced back to a period, when the refilling of the east lobe lead to an episodic formation of a freshwater lense, whereby an ice cover could be established, but was non-perennial. Contemporaneously, salt accumulation continued, which suggests the brine remained supersaturated with respect to sodium and chloride either by cryo-concentration of freshwater or inflow of saline waters.

Differing properties of unit 4 compared to units 1 and 3 of core Lz1023 suggest a significant change in environmental conditions. High amounts of clastic material in unit 4 (Figure 24) indicate dissolution of halite or an interruption of halite precipitation. Furthermore, the occurrence of carbonates and gypsum in the upper part of core Lz1023 (Figure 24) point to a reversal of the depositional cycle of evaporites. Both can be associated with a freshwater input to the lake. Hendy et al. (1977; 1979) also suggest that halite precipitation has been interrupted in East Lake Bonney in the recent past and document a reversal of the depositional cycle of evaporites in form of aragonite and gypsum precipitates in the surface sediments of ELB. U/Th dating on these minerals yielded ages between 100 and 2000 years (Table 7). In addition, diffusion gradients in Taylor Valley lakes indicate a refilling from 1000 years BP (Lyons et al., 1998). After Poreda et al. (2004), East Lake Bonney was successively refilled, until it finally reached the same level as the west lobe, and remained ice-free until 200 years ago (Figure 27). At this time, a perennial ice cover was formed in the east lobe for the first time, since the high salinity of the water prohibited this in former times. In core Lz1023, a perennial ice cover is indicated by the dominance of silt and lack of gravel in unit 4 (Figure 24). With the establishment of a permanent freshwater lense at East Lake Bonney, a perennial ice cover was formed, possibly only a few hundred years ago, prohibiting coarse-grained material to deposit in the lake. Diffusion processes occurring higher in the water column today (Spigel and Priscu, 1996) and the observation of a rising lake level during the last century (Chinn, 1993) support the assumption of elevated freshwater input to Lake Bonney and that precipitation of halite from the brine is currently interrupted.

5.6 Conclusions

Sedimentological, biogeochemical and chronological investigations on a 2.7 m long sediment core (Lz1023) recovered from East Lake Bonney, Taylor Valley, southern Victoria Land, Antarctica, provide paleoenvironmental information about the lake's history. The core mainly consists of halite crystals building up a more than 2 m thick salt crust. Although several hypersaline lakes are known from Antarctica (Matsubaya et al., 1979), to our knowledge, halite deposits in this thickness was hitherto only recovered from East Lake Bonney. The salt crust mainly composed of halite crystals shows two cyclic successions (units 1 and 3) with distinct variations in crystal sizes. These successions are interrupted by a horizon (unit 2), whose properties are differing from the majority of the core and which either consists of gashydrates or metastable minerals. The uppermost 60 cm of the core can be classified as transition between salt crust and water column (unit 4).

Since radiocarbon dating on the sediments provided a questionable age, variations in the sedimentological composition can only be used for first paleoenvironmental implications and deducing signs of environmental change. Even if radiocarbon dating yielded an age of ~11,000 ^{14}C years at 172 cm depth, high reservoir effects in the lake's bottom waters suggest an younger age. The assumed continuous sedimentation rate based on constant incorporation of clastic material in units 1 to 3 imply that the salt crust was formed within few thousand years, most likely during the Late Holocene. This is in accordance with investigations based on element concentration gradients and isotopes in the present water column (e.g., Hendy et al., 1977; Lyons et al., 1999; Poreda et al., 2004). While the origin of hypersaline water in the eastern lobe is not definitely clear, its supersaturation with respect to sodium and chloride may have lead to the precipitation of halite during a cold and dry period between 2000 and 1000 years BP (Lyons et al., 1998). Variations in crystal sizes and occurrence of carbonate and gypsum layers indicate changes in the constitution of the precipitating brine and in water supply. However, the occurrence of gravels in the upper part of unit 3 can be traced back to episodic ice formation on the lake, which implies the establishment of a freshwater lens in East Lake Bonney for longer periods. Refilling of the basin with freshwater was probably responsible for the interruption of halite precipitation and for the occurrence of carbonates and gypsum layers in unit 4. Fine-grained clastic material and lacking gravels suggest the establishment of a perennial ice cover in the recent past.

The sediment sequence from East Lake Bonney (Lz1023) can be regarded as a unique record, even if its utility for paleoenvironmental reconstruction is limited.

6 Lacustrine history of Taylor Valley - Synthesis

6.1 Introduction

Taylor Valley covers a relatively small area of approximately 400 km^2 (Figure 28). The three major lakes of this valley, lakes Fryxell, Hoare and Bonney, show differences in water chemistry, ecology, and also in their sediments. This is remarkable, because these lakes represent all remnants of proglacial Lake Washburn, which occupied Taylor Valley during the last glacial period (Hall et al., 2000). After the desiccation of Lake Washburn, their location within the valley and their individual histories created the present characters of the Taylor Valley lakes (Lyons et al., 2000). Especially, accumulation of salts over time, for instance in Lake Bonney, changed the composition of the water column and the character of the sediments. Due to their location within Taylor Valley, the lakes are influenced by the different factors, mainly by their proximity to the sea, the type of the surrounding landscape, geomorphology, and adjacencies to glaciers (Lyons et al., 2000).

Figure 28. Taylor Valley. (a) Location of McMurdo Dry Valleys in Antarctica (U.S. Geological Survey, 2006). (b) Landsat-7 satellite image of Taylor Valley, McMurdo Dry Valleys, Antarctica, indicating the location of the most important lakes and glaciers (U.S. Geological Survey, 2007).

The proximity to the sea has an influence on microclimatic conditions (c.f. Fountain et al., 1999). The type of the surrounding landscape plays an important role regarding the differences in sedimentology between lakes Fryxell and Hoare. Lake Fryxell is located in a broad and flat part of the valley floor, whereas Lake Hoare lies in a narrow part with steep slopes (Figure 28). While sediment input to Lake Fryxell is more dependent on streams, sediment supply to Lake Hoare is affected by the slopes and the wind blocking effect of Canada Glacier (Doran et al., 1994). This causes the deposition of the high amounts of material on the ice cover and results in an overall coarser grain size of Lake Hoare sediments compared to Lake Fryxell sediments.

The general geomorphology of the valley floor forming the bathymetries of the lakes has an impact on the lakes' water chemistry and sedimentology. This can be best illustrated with the help of Lake Bonney as an example. This lake is divided into two lobes by a subaquatic sill, what causes differences in water chemistry and sedimentology of the eastern and western lobe (Lyons et al., 2005). Another aspect illustrates the importance of the adjacencies to glaciers, which affects the mode of meltwater supply (directly or via streams) and the solute composition of the lake water. The western lobe of Lake Bonney is located directly at the tongue of Taylor Glacier, which is also advancing into the lake (Higgins et al., 2000b). Today, meltwater deriving from this glacier feeds the lake and is jointly responsible for the current lake level rise (Chinn, 1993). A special feature of the Taylor Glacier is the Blood falls, bringing highly saline and iron-rich waters into West Lake Bonney (Mikucki et al., 2004). Likewise, Lake Hoare is directly connected to Canada Glacier and its meltwater supply. The lake water is characterized by low ion concentrations, since its waters mainly derive from Canada Glacier melt (Fountain et al., 1998).

According to all this, several factors have an influence on the hydrology, physical properties, water chemistry, and ecology of the lakes of Taylor Valley. The resulting differences have consequences on the characteristics of the lake sediments and have to be taken into account for paleoenvironmental reconstructions. Furthermore, the investigations on the Taylor Valley lake sediment records reveal that only a multidisciplinary approach allows conclusions on past environments and climates.

6.2 Assessment of proxies for paleoenvironmental reconstructions

The sedimentological, biogeochemical, and mineralogical investigations on the lake sediment cores of Taylor Valley show that certain proxies proved to be particularly relevant for the reconstructions within this study. Volcanic glass, for instance, is preserved in the lake sediments of Lake Hoare (chapter 4, Figure 14), from which the influence of an advanced RIS at the mouth of Taylor Valley can be inferred (c.f. Denton and Hughes, 2000). Other mineralogical data turned out to be less informative, what can be traced back to the geology of the region. The mineralogical composition of the geological units in the catchment is very similar due to crystalline basement with dominance of feldspars, quartz, and pyroxenes. Amphiboles show a correlation to the volcanic glass contents, which points to basaltic rocks of Ross Island as a common source area (c.f. Wagner et al., 2006). Slight variations in the mineralogy of the sediments refer to changes in the source area on the one hand, and to different depositional processes on the other hand. As shown by the Lake Hoare record (chapter 4, Figure 14), the advance of Canda Glacier during the Mid-Holocene was responsible for an increased feldspar supply to the lake.

The grain-size distribution of the lake sediments can be used to reconstruct transport and depositional processes. Sandy and gravelly sediments dominate the valley floor today, since the strong winds deflate the fine-grained material. However, sand can also be transported aeolianly. The dominance of sands and silts in Taylor Valley lake sediments (chapter 4, Figure 14; chapter 5, Figure 24) can be traced back to aeolian and (glacio-)fluviatile sources. However, the dominant mode of deposition is controlled by the ice cover. Clastic material is trapped on and within the lake ice. This material can migrate by freezing and thawing processes through the perennial ice cover or can fall through cracks (Nedell et al., 1987). Sand mounts presently found at the bottom of Lake Hoare indicate a highly variable sediment deposition, which is mainly controlled by the ice cover (Squyres et al., 1991). The occurrence of gravels indicates a change in ice cover dynamics, since under present conditions grains larger than 1 cm in diameter cannot melt through the perennial ice cover (Hendy, 2000b). Gravels occurring in the upper part of unit 3 of core Lz1023 (chapter 5, Figure 24) recovered from East Lake Bonney point to the formation of floating ice on the lake. In the Lake Hoare record (chapter 4, Figure 17), periods with a non-perennial ice cover of glacial Lake Washburn can be deduced by the occurrence of gravel in the sediments. Furthermore, gravelly sediments in a lacustrine environment are mainly restricted to the lake edges and can be accumulated in high amounts during lake level drops, because larger clasts remain and accumulate on the ice cover, and are released to the lake bottom not until the ice cover is melted (Hall et al., 2006a; Hendy et al., 2000). For instance, unit III of core Lz1020 (chapter 4, Figure 17) recovered from Lake Hoare shows indications for a desiccation of the coring site, since large gravel clasts can be regarded as lag deposits.

Concerning biogeochemical proxies, the contents of inorganic carbon (TIC) as well as of total sulfur (TS) are useful indicators for lake water conditions. Precipitation of carbonates, which is reflected in high TIC values, results from higher ion concentrations in the water column. This in turn can be a consequence of lake level lowering due to evaporation, since the lakes had or have no outflow. For instance, the final evaporation of glacial Lake Washburn is indicated by an increase in TIC in unit II of the Lake Hoare record (chapter 4, Figure 14). High contents of sulfur in the sediment can be traced back to anoxic bottom waters in the lake, as observable in the upper part of unit IV of the Lake Hoare record (chapter 4, Figure 14). Present investigations show that sulfur is fixed in surface sediments by formation of pyrite (Bishop et al., 2001). In an evaporite environment, TS can also reflect the occurrence of gypsum in the sediment as shown by investigations on the East Lake Bonney record (chapter 5). Total organic carbon (TOC) contents in Taylor Valley lake sediments are very low and variations in TOC are less significant (chapter 4, Figure 14; chapter 5, Figure 24), what can be related to the low bioproductivity in Antarctic lakes and the hindered preservation of organic matter, respectively (Hodgson et al., 2004).

Challenges evoke from dating the sediments of Lake Hoare (core Lz1020) and East Lake Bonney (core Lz1023), since radiocarbon dating of Antarctic sediments includes diverse difficulties. Large reservoir effects of several thousand years, as assumed for bottom waters of East Lake Bonney (chapter 5, Figure 26) are common, since the perennial ice cover prohibits an exchange of the water column with the atmosphere (Hendy and Hall, 2006). Additionally, the input of old carbon from glacial melt and glacio-lacustrine sediments results in ages older than the actual deposition time. The recent reservoir effect in Lake Hoare could be estimated by dating the surface sediments and amounts to over 4000 years (chapter 4, Table 4). In addition, as explicitly shown for the Lake Hoare record (chapter 4, Figure 16), the reservoir effect in the lake has changed in the past. These uncertainties make it difficult to create a reliable chronology. However, concerning Lake Hoare sediments, other methods are non-applicable, like U/Th dating, when sufficient carbonates are lacking, or they are less accurate, like luminescence dating (Doran et al., 1999). Otherwise, the U/Th method could help dating East Lake Bonney sediments, but could not be carried out within the scope of this study.

6.3 Reconstruction of the late Quaternary environmental history of Taylor Valley, Antarctica

Lake sediment records of the three major lakes Fryxell, Hoare and Bonney (Figure 28) all provide information about the late Quaternary environmental history of Taylor Valley, but due to their differing characteristics, these records are different in their significance as archives.

Lake Fryxell provides the best location for longterm records, since it is situated in the deepest part of the valley, in a potential accumulation area (Figure 28). Wagner (2003) recovered a nearly 10 m long core from the deepest part of Lake Fryxell, which covers the last 48,000 years and indicates lacustrine conditions at the coring site throughout all the time (Figure 29).

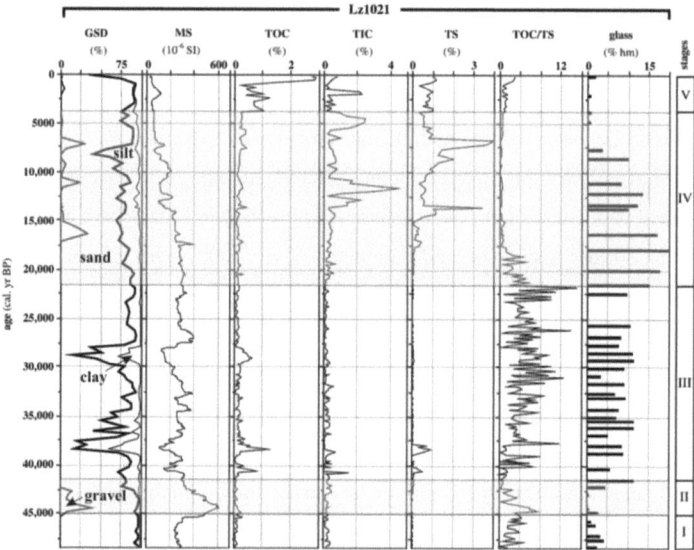

Figure 29. The Lake Fryxell record (modified after Wagner et al., 2006) with grain-size distribution (GSD), magnetic suszeptibility (MS), total organic carbon (TOC), total inorganic carbon (TIC), total sulfur (TS), TOC/TS ratio and amounts of volcanic glass in the heavy mineral fraction plotted versus the calibrated radiocarbon age. The differentiated stages refer to periods of lake evolution of Lake Fryxell.

Between 48,000 and 42,000 cal. years BP (stages I and II in Figure 29), the basin of eastern Taylor Valley was likely occupied by a small lake. From 42,000 to 8000 cal. years BP (stage III and IV in Figure 29), high amounts of volcanic glass in the lake sediments suggest that an advanced Ross Sea Ice Sheet (RIS) blocked the valley mouth and dammed the large proglacial Lake Washburn (Wagner et al., 2006). Geomorphological evidence indicates the occupation of Taylor Valley by Lake Washburn between 23,800 and 8340 ^{14}C years BP (Hall and Denton, 2000b) (Figure 32). The existence of the large proglacial lake during generally very cold and dry climate is explained by Hendy (2000a) with an elevated meltwater supply from the RIS due to the higher numbers of clear summer days.

The Lake Hoare record (chapter 4) shows that the coring site of Lz1020 was also occupied by Lake Washburn during the last glacial period. Coarse sediments with gravels in unit I (chapter 4, Figure 14) indicate that between 17,000 and 14,000 years BP Lake Washburn had a non-perennial ice cover (Figure 30a). The proposed high lake level during this period (Figure 30a) suggests a high meltwater supply, possibly due to climate warming between 16,000 to 14,000 years BP, following the Last Glacial Maximum (Steig et al., 2000). The occupation of western Taylor Valley by Lake Washburn is confirmed by the occurrence of paleodeltas dating from 23,800 to 11,800 ^{14}C years BP (Hall and Denton, 2000b). Reconstruction with the help of the lake sediments recovered from ELB are not possible, since the record (core Lz1023) does not penetrate back to this period.

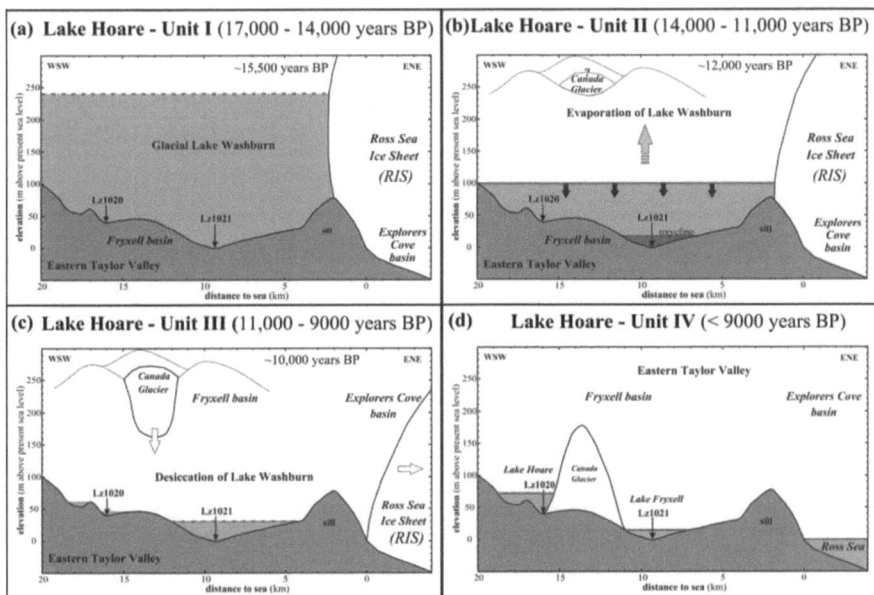

Figure 30. Reconstruction of the late Quaternary environmental history of eastern Taylor Valley. Scenarios for the profile of eastern Taylor Valley (in fortyfold vertical exaggeration) are shown for particular times within the different stages of lake history, based on sediment units of the Lake Hoare record (chapter 4, Figure 14) and published data mentioned in the text.

Environmental and climate conditions changed during the Pleistocene-Holocene transition. Increasing TIC contents in unit II of the Lake Hoare sequence (chapter 4, Figure 14), also observable in the Lake Fryxell record (stage IV in Figure 29), suggest a gradual evaporation of Lake Washburn between 14,000 and 11,000 years BP (Figure 30b). Lake level lowering during this period is confirmed by decreasing elevations of paleodeltas both in eastern and western Taylor Valley (Hall and Denton, 2000b). The RIS still occupied the valley mouth (Figure 30b), as indicated by the occurrence of volcanic glass and elevated amounts of amphiboles in unit II of the Lake Hoare record (chapter 4, Figure 14). In addition, the fine grain sizes of the Lake Hoare sediments in unit II (chapter 4, Figure 17) suggest that Lake Washburn had a perennial ice cover (Figure 30b). Indications for anoxic bottom water conditions by elevated TS contents in the Lake Fryxell record (stage IV in Figure 29) support the supposition of a perennial ice cover during this period. The sediments of Lake Fryxell and Hoare possibly originate from a cooling period between 14,000 and 12,000 years BP, probably correlated to the Younger Dryas stade in the Northern Hemisphere (Steig et al., 2000). This period was responsible for cooler and drier climate conditions in the dry valleys, whereas evaporation processes were intensified. Consequently, evaporation of the proglacial lakes in the McMurdo Dry Valleys is also documented from 14,000 ^{14}C years BP (Figure 32).

Lag deposits in unit III of the Lake Hoare record (core Lz1020) indicate a desiccation of the coring site during the period between ~11,000 and 9000 years BP (Figure 30c). Enhanced evaporation was likely responsible for a further lake level drop of Lake Washburn, as also suggested by carbonate precipitation in the Lake Fryxell record (Wagner et al., 2006) (stage IV in Figure 29). This is also confirmed by other studies (e.g., Lawrence and Hendy, 1989; Whittaker et al., 2008), mostly in form of carbonate precipitation in lakes dating from 11,000-10,000 years BP (Figure 32). Evidence for an outflow event of Lake Washburn could be found neither in our data nor in geomorphological investigations in Taylor Valley. However, the Taylor Dome ice core record shows a trend to a warmer and more humid climate (Steig et al., 2000). Probably, the warming during the Pleistocene-Holocene transition was responsible for enhanced evaporation in the region. As the coring site of core Lz1020 is located at 40 m a.s.l. and fell dry between 11,000 and 9000 years BP, it can be assumed that from this time, Lake Washburn was divided into two remnant lakes, which occupied the eastern Fryxell basin and the western Bonney basin separately. The separation of Lake Washburn into two basins is also confirmed by the occurrence of paleodeltas at elevations below the threshold between the Bonney and Fryxell basin (116 m a.s.l., red line in Figure 31) from ca. 11,000 ^{14}C years BP (Hall and Denton, 2000b). Characteristics of unit III of the Lake Hoare record (Figure 30c) indicate a drastic lake level drop in eastern Taylor Valley until ca. 9000 years BP due to evaporation. Thus, it can be assumed that the remnant of Lake Washburn in the western Bonney basin has also evaporated to at least a very small volume, probably forming small ponds in the depressions. Even though there is no clear evidence from the East Lake Bonney record (core Lz1023) for a permanent existence of a lake in the basin during the whole Holocene, based on the high salinity of the present brine it can be assumed that a hypersaline brine could have persisted in the basin (Krumgalz, 2001). Furthermore, low lake levels (<40 m a.s.l.) in eastern Taylor Valley, as indicated by unit III of the Lake Hoare record, suggest that the remnant of glacial Lake Washburn in eastern Taylor Valley lost its connection to the RIS (Figure 30c), since the lake level also dropped below the threshold between the Fryxell basin and Explorers Cove (75 m a.s.l., blue line in Figure 31) According to this, meltwater supply from the RIS would have been interrupted. Otherwise, evidence for volcanic glass in the sediments of lakes Hoare and Fryxell (chapter 4, Figure 18) suggest that the Ross Sea Ice Sheet likely occupied the mouth of Taylor Valley until ca. 8000 cal. years BP (Figure 32). Erosion of older Lake Washburn sediments from the valley slopes could explain the sparse occurrence of volcanic glass in unit III of the Lake Hoare record (chapter 4, Figure 14), in times when its supply by the RIS can be widely excluded.

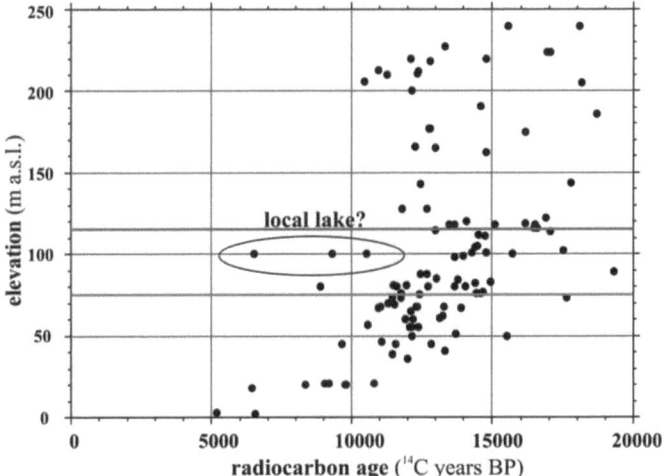

Figure 31. Radiocarbon ages of organic remains from ancient deltas versus their altitude in Taylor Valley (data from Hall and Denton, 2000b).

Geomorphological evidence show that the RIS retreated from Explorers Cove between 8340 and 6500 ^{14}C years BP (Hall and Denton, 2000b). Since unit IV of the Lake Hoare record shows no significant changes of its sediments (chapter 4, Figure 14), a remnant of Lake Washburn occupied eastern Taylor Valley, probably in its characteristics similar to the modern lake. Only mineralogical data show indications for a change in source areas of lake sediments. Increasing amounts of feldspars in the upper part of unit IV of the Lake Hoare record imply that the Canada Glacier advanced during the Mid-Holocene (around 6000 years BP), finally separating lakes Hoare and Fryxell (Figure 30d). This can be associated to the Alpine I drift (Figure 32). Taylor Glacier and alpine glaciers in Taylor Valley advanced, because the regional climate became more humid during the Holocene with the retreat of the RIS from McMurdo Sound (Steig et al., 2000). In addition, the Mid-Holocene temperature rise was responsible for warmer and moister climate conditions in the region. According to this, source of water for Taylor Valley lakes have mainly derived from meltwater of alpine glaciers during the Holocene. Refilling of the Lake Hoare basin with fresh meltwater of Canada Glacier is indicated by lack of TIC in the upper part of unit IV (chapter 4, Figure 14). After Hendy et al. (1977), Taylor Glacier advance in western Taylor Valley was responsible for the refilling of West Lake Bonney. The basin of East Lake Bonney is said to had been occupied by a stagnant water body (Poreda et al., 2004). Reconstructions with the help of the lake sediments recovered from ELB are not possible, since our record (core Lz1023) does not penetrate back to this time.

Variations in the regional Holocene climate can be deduced from lake level fluctuations. Especially, cold and dry periods are related to evaporation events in the dry valleys (Hendy, 2000a). Two significant carbonate precipitation events are indicated in the Lake Fryxell record at ~5000 cal. years BP and ~2000 cal. years BP (Wagner et al., 2006) (Figure 29). Latest investigations by Whittaker et al. (2008) detected evaporation events in Lake Fryxell dating from 6400, 4700, 3800 and 1600 cal. years BP and assume that they were of minor extent (<4.5 m below the present lake level). Due to low or lacking TIC contents in its sediments (chapter 4, Figure 14), reconstructions of lake level oscillations based on the Lake Hoare record are difficult. Indications for anoxic bottom waters in Lake Hoare by increasing TS contents in the upper part of unit IV (chapter 4, Figure 14) could be related to a period with high lake levels in the dry valleys between 3000 and 2000 years BP (Figure 32). Signs for a complete desiccation of Lake Hoare afterwards, as suggested by Lyons et al. (1998), could not be found in the record of core Lz1020. Probably, lake level lowering also was of minor extent, since sampling site DH 2 at 11 m water depth were occupied by the lake at least during the last 2500 years (see Doran et al., 1999). In this context, evaporites (units 1-3 in core Lz1023) found in East Lake Bonney (chapter 5, Figure 24) confirm the occurrence of an evaporation event in the late Holocene. Dry and cold climate conditions between 2000 and 1000 years BP are likely responsible for the accumulation of the halite crust at the bottom of East Lake Bonney (Figure 32). The occurrence of this second significant evaporation event, that followed a lake level highstand, is suggested in all of the three Taylor Valley lake sediment records (Figure 32) and can also be found in other dry valley lakes (e.g., Smith and Friedman, 1993). For these reasons, it seems to be at least a regional event, indicating a dry and cold period between 2000 and 1000 years BP (Figure 32). The records of Lake Hoare and East Lake Bonney imply that following this dry period, lake levels have been rising in the last ~1000 years. The establishment of anoxic bottom water conditions in Lake Hoare due to a higher lake level is indicated by enhanced TS contents in the uppermost part of unit IV (chapter 4, Figure 14). Refreshening processes in East Lake Bonney are inferable from characteristics of unit 4 of core Lz1023 (chapter 5, Figure 24), where an interruption of halite accumulation can be observed and precipitation of gypsum and carbonates indicate a reversal of the evaporite deposition cycle. Lyons et al. (1998) ascribe this phenomenon of lake level rising in Taylor Valley to relatively mild climate conditions. When refilling of East Lake Bonney basin reached the same level like as West Lake Bonney (chapter 5, Figure 27), a freshwater lens was established above the whole lake. According to this, East Lake Bonney may have formed a permanent ice cover only few hundred years ago, whereas lakes Hoare and Fryxell have been perennial ice-covered throughout most of the Holocene (Figure 31d).

6 Lacustrine history of Taylor Valley

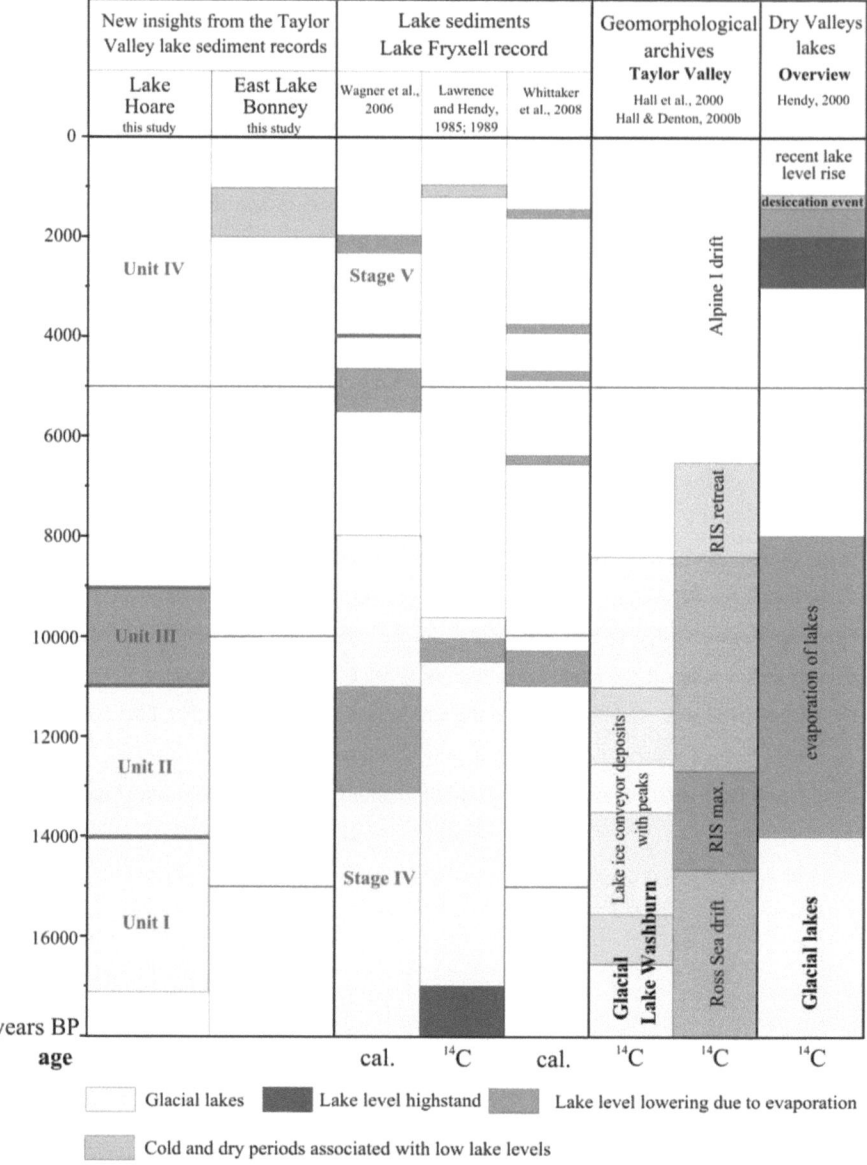

Figure 32. Overview of the late Quaternary environmental history of Taylor Valley. New insights about the environmental history of Taylor Valley are inferred from the Lake Hoare (core Lz1020) and East Lake Bonney (Lz1023) records and are compared to information about the environmental history, which is summarized from the most important studies mentioned in the text and includes geomorphological archives and lake sediment records of the dry valleys.

6.4 Implications for the Holocene climate history of East Antarctica

Climate reconstructions based on the lake sediment records alone are difficult. Although lake level changes are mainly caused by climatic variations, local influences are also reflected in the records. Nevertheless, lake sediment records of Taylor Valley not only provide new information about the environmental history of the region, but also contribute to the understanding of the Holocene climate history of East Antarctica.

Whereas ice core records from continental Antarctica show less temperature variations throughout the Holocene (Vostok in Figure 33), ice core records from the Ross Sector (Taylor Dome and Byrd in Figure 33) imply an alternation of warmer and cooler periods. Sedimentary records from East Antarctica (Figure 33) also show indications for climate changes during the Holocene, even if the number and timing of warm periods are varying. On the one hand, this can be traced back to dating uncertainties. On the other hand, the documentation of climate changes in the sediment records are locally modified.

Most of the East Antarctic oases were glaciated during the Last Glacial Maximum (LGM). The timing of deglaciation in East Antarctica was variable and dependent on local conditions (Figure 33). In McMurdo Dry Valleys, the deglaciation history has to be seen in terms of the RIS dynamics, since the dry valleys have been mostly ice-free at least for the Quaternary. During the LGM, large parts of the Ross Sea were occupied by a thick ice sheet, which was also responsible for the damming of proglacial lakes in the dry valleys (Hendy, 2000a). The climate warming during the Pleistocene-Holocene transition was responsible for the retreat of the Ross Sea ice sheet (RIS). Volcanic glass occurrence until ~8000 cal. years BP in the Lake Fryxell record (Wagner et al., 2006) implies that the RIS had occupied the mouth of Taylor Valley until this time (Figure 33). Furthermore, warming during the Pleistocene-Holocene transition may have lead to the evaporation of the large glacial lakes in the McMurdo Dry Valleys. This is supported by lag deposits in unit III of the Lake Hoare record (chapter 4, Figure 17), which points to a desiccation of Lake Washburn in Taylor Valley between 11,000 and 9000 years BP (Figure 33). Masson et al. (2000) define this period as early Holocene climate optimum in the Ross Sector. Indications for this climate amelioration can also be found in sediment records of the Larsemann Hills between ca. 11,500 and 9500 cal. years BP (Verleyen et al., 2004) (Figure 33). Warm periods in the early Holocene documented in Untersee, Amery and Bunger Oases around 9000 years BP (Kulbe et al., 2001; Schwab, 1998; Wagner et al., 2004) cannot be found directly in the Taylor Valley lake sediment sequences (Figure 33). However, signs for a continuing warming in the region can be deduced indirectly from indications for the final retreat of the RIS from the mouth of Taylor Valley between 8340 and 6500 ^{14}C years BP (Hall and Denton, 2000b).

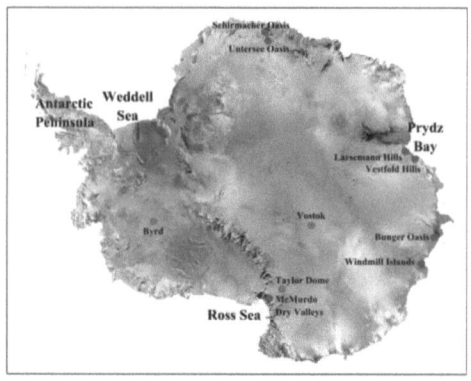

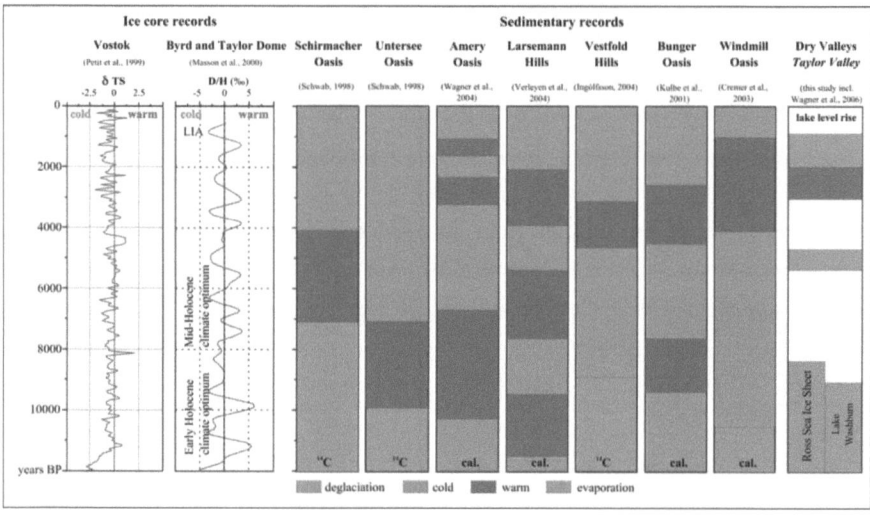

Figure 33. Holocene climate history of East Antarctica inferred from different archives. The figure summarizes important paleoclimate data of ice core records (Petit et al., 1999; Masson et al., 2000) and sedimentary records of East Antarctic Oases (Schwab, 1998; Kulbe et al., 2001; Cremer et al., 2003; Verleyen et al., 2004; Ingólfsson, 2004; Wagner et al., 2004) including new information obtained from Taylor Valley in McMurdo Dry Valleys. Locations of the oases (red points) and ice cores (blue points) are shown in the map above.

Known from other studies in McMurdo Dry Valleys (e.g., Smith and Friedman, 1993), cooler periods in the region are characterized by drier climate conditions, which are responsible for evaporation events and lower lake levels. An evaporation event in Taylor Valley, as indicated by the Lake Fryxell record, around 5000 cal. years BP seems to coincide with a cooler period following the Mid-Holocene climate optimum after Masson et al. (2000).

Except for the Schirmacher Oasis, cooler climate is also documented in the other East Antarctic Oases (Figure 33). Studies in the Larsemann Hills additionally show that lakes had a high level during wet and warm climate conditions, especially in two significant periods, at ca. 4000 cal. years BP and between 2500 and 3000 cal. years BP (Verleyen et al., 2004). Indications for high lake levels during the latter period can also be found in lake sediments of eastern Taylor Valley (Figure 33). The period with warmer climate in East Antarctica in the second half of the Holocene (Ingólfsson, 2004), as observable in several oases (Figure 33), is discussed as Mid-Holocene Hypsithermal and occurs throughout Antarctica between 5000 and 2000 years BP (Bentley and Hodgson, 2009). In Taylor Valley, this warm period with high lake levels was followed by a significant evaporation period (Figure 33), best illustrated by the thick salt deposits in East Lake Bonney. Colder temperatures coupled with dryness led to evaporation, most likely between 2000 and 1000 years BP. Masson et al. (2000) deduce a cold period from Taylor Dome and Byrd ice core records at ca. 1000 years BP, which they interpret as Antarctic Little Ice Age (LIA in Figure 33). Due to the dating uncertainties in our cores, it is quite possible that the late Holocene evaporation event occurring in core Lz1023 from East Lake Bonney correlates with the LIA. Lake level rise in the recent past could be traced back to a shift to moister conditions, which indicates a slight warming after the LIA. However, this is not necessarily related to a real warm period, since ice core records show a general cooling trend in the late Holocene (Figure 33), discussed as Neoglacial cooling (Ingólfsson, 2004).

6.5 Conclusions

The cores obtained from lakes Fryxell (Lz1021), Hoare (Lz1020), and East Lake Bonney (Lz1023) are the longest lake sediment sequences yet recovered from the Taylor Valley lakes, but are varying in their significance as archives due to their different characteristics. Their sedimentological variability is remarkable, considering the three lakes are located within a relatively small area, the same valley, and have a "shared" past as remnants of the large glacial Lake Washburn. Whilst Lake Fryxell sediments (core Lz1021) penetrates back into the late Pleistocene to 48,000 cal. years BP and provides information about lake level fluctuations in the past (c.f. Wagner et al., 2006), the Lake Hoare sediment sequence (core Lz1020) is limited in its significance due to the dominance of coarse clastic material and covers the last 17,000 years. The specific feature of core Lz1023 recovered from East Lake Bonney is the more than 2 m thick salt deposit, which likely originate from the late Holocene.

The investigations on the Taylor Valley lake sediment records reveal that only a multidisciplinary approach allows conclusions on past environments and climates. Furthermore, interpretation of the lake sediment records requires good knowledge about the environment of the lake and its changes over time. In the current study, available information from geomorphological investigations facilitates the interpretation of the data obtained from the lake sediment sequences. In spite of that, the records give new insights, where geomorphological information is limited due to its poorer preservation ability.

This highlights that lake sediment sequences of Taylor Valley can provide crucial information about the regional environmental history, but should be interpreted in context with findings from other records, like ice core, marine, and terrestrial records. Nevertheless, the integration of the paleoenvironmental information deduced from the individual lake sediment records not only allows a detailed reconstruction of the Late Quaternary environmental history of Taylor Valley, but also contributes to the understanding of the late Quaternary climate history of East Antarctica.

7 Summary and final remarks

In this study, lake sediment cores recovered from Lake Hoare (Lz1020) and East Lake Bonney (Lz1023) were investigated due to their sedimentological, biogeochemical, mineralogical properties, and their chronology. This multidisciplinary approach provided new crucial information about the late Quaternary regional environmental history of Taylor Valley, southern Victoria Land, Antarctica.

Sediment cores from Lake Hoare penetrate back into the late Weichselian, to ~17,000 years BP, when Taylor Valley was occupied by the large proglacial Lake Washburn. The occurrence of volcanic glass in the sediments confirms that this lake was dammed by the advanced Ross Sea Ice Sheet at the valley outlet. During the Pleistocene-Holocene transition, between ~14,000 and 11,000 years BP, enhanced evaporation led to a significant lake level drop of Lake Washburn. This is indicated by elevated amounts of inorganic carbon in the sediments. The Lake Hoare record additionally shows that between 11,000 and 9,000 years BP, Lake Washburn desiccated to a very low level, with subaerial conditions at the coring site of Lz1020. After the final retreat of the Ross Sea Ice Sheet during the early Holocene, Taylor Valley was occupied by remnants of Lake Washburn. Changes in the mineralogical composition and sedimentological differences of the Lake Hoare sediments compared to Lake Fryxell sediments indicate that environmental conditions comparable to those of today, with an advanced Canada Glacier separating lakes Hoare and Fryxell, established during the Mid-Holocene.

The most important insight provided by the East Lake Bonney record is that the central part of the basin contains a more than 2 m thick salt crust. The sediment is predominately composed of halite, which corroborates the longer-term constitution of hypersaline bottom waters in East Lake Bonney with supersaturation with respect to sodium and chloride. It was shown, that between ~2000 and 1000 years BP, a significant evaporation event led to the deposition of a thick salt crust in the basin of East Lake Bonney. Following this event, changing climate conditions in the late Holocene are responsible for a lake level rise, which is reflected in the sediments by a reversal of the evaporite depositional cycle. The establishment of a perennial ice cover in the recent past is confirmed by the lack of gravels in the clastic fraction.

The establishment of an independent and reliable chronology based on the lake sediment sequences with the help of radiocarbon dating turned out to be difficult due to changing reservoir effects in the lakes' past and glacial contaminations. However, the sediment sequence recovered from Lake Hoare (Lz1020) provides a long record continuously covering the Holocene and penetrating back into the late Pleistocene. Even if the Lake Hoare sediment sequence is limited in its significance due to the dominance of coarse clastic material and chronological uncertainties, the record provides new information on the desiccation of Lake Washburn and the separation of lakes Hoare and Fryxell in the Mid-Holocene by the advance of Canada Glacier. The East Lake Bonney record shows indications for enhanced evaporation in the late Holocene and for a lake level rise in the recent past.

The investigations on the lake sediment records contribute to a better understanding of ocean-land-ice-interactions, as the late Quaternary history of Taylor Valley lakes is highly dependent on climate changes, dynamics of the Ross Sea ice sheet and local glacier systems. The new insights provided by the lake sediment records broaden the understanding of the late Quaternary environmental history of Taylor Valley, if they are interpreted involving information deduced from other archives, like ice-core, marine, and terrestrial records.

Since the McMurdo Dry Valleys region was ice-free during the last glacial, the specialty of lake sediments of Taylor Valley can be seen in providing longer records compared to most other Antarctic oases. Based on the lake sediment cores of Lake Hoare and in correlation with the Lake Fryxell record, the late Quaternary environmental history of eastern Taylor Valley can be very well reconstructed. However, information about the western Taylor Valley can only be given for the late Holocene by the East Lake Bonney record. The coring of long sequences down to and into the base of the evaporites could provide long-term records and additional information on the history of western Taylor Valley. Furthermore, East Lake Bonney has a complicated system of deposition, whereby inferring of paleoenvironmental information turned out to be difficult. This highlights the necessity of a process-analytical approach in the study of lake sediments. Paleoenvironmental reconstructions could be improved, if the processes of deposition were understood in detail. These gaps in our knowledge require further studies on geochemistry, fluid inclusions, and stable isotopes of the salts in East Lake Bonney.

References

Andersen, D.W., Wharton jr., R.A., and Squyres, S.W., 1993, Terrigenous clastic sedimentation in Antarctic Dry Valley lakes. Physical and biogeochemical processes in Antarctic lakes: Antarctic Research Series, v. 59, p. 71-81.

Angino, E.E., Turner, M.D., and Zeller, E.J., 1962, Reconnaissance geology of the lower Taylor Valley, Victoria Land, Antarctica. : Geological Society of America Bulletin, v. 73, p. 1553-1562.

Baroni, C., and Orombelli, G., 1994a, Abandoned penguin rookeries as Holocene paleoclimatic indicators in Antarctica: Geology, v. 22, p. 23-26.

Baroni, C., and Orombelli, G., 1994b, Holocene glacier variations in the Terra Nova Bay area (Victoria Land, Antarctica): Antarctic Science, v. 6, p. 497-505.

Bentley, M.J., and Hodgson, D.A., 2009, Antarctic Ice Sheet and climate history since the Last Glacial Maximum: PAGES News, v. 17, p. 28-29.

Berkman, P.A., Andrews, J.T., Björck, S., Colhoun, E.A., Emslie, S.D., Goodwin, I.A., Hall, B.L., Hart, C.P., Hirakawa, K., Igarashi, A., Ingólfsson, O., López-Martínez, J., Lyons, W.B., Mabin, M.C.G., Quilty, P.G., Taviani, M., and Yoshida, Y., 1998, Circum-Antarctic coastal environmental shifts during the Late Quaternary reflected by emerged marine deposits: Antarctic Science, v. 10, p. 345-362.

Bindschadler, R., 1998, Future of the West Antarctic Ice Sheet: Science, v. 282, p. 428-429.

Bishop, J.L., Lougear, A., Newton, J., Doran, P.T., Froeschl, H., Trautwein, A.X., Körner, W., and Koeberl, C., 2001, Mineralocical and geochemical analysis of Antarctic lake sediments: A study of reflectance and Mössbauer spectroscopy and C, N, and S isotopes with applications for remote sensing on Mars: Geochimica et Cosmochimica Acta, v. 65 p. 2875 - 2897.

Bromley, A.M., 1985, Weather observations, Wright Valley, Antarctica: Wellington, 37 p.

Burkemper, A., 2007, Lacustrine history of Lake Hoare in Taylor Valley, Antarctica, based on long sediment cores [Master thesis]: Chicago, University of Illinois.

Chinn, T.J., 1993, Physical hydrology of the dry valley lakes, in Green, W.J., and Friedman, G.M., eds., Physical and biological processes in Antarctic lakes: Antarctic Research Series: Washington D.C., p. 1-51.

Clayton-Greene, J.M., Hendy, C.H., and Hogg, A.G., 1988, Chronology of a Wisconsin age proglacial lake in the Miers Valley, Antarctica: New Zealand journal of Geology and Geophysics, v. 31, p. 353-361.

Clow, G.D., McKay, C.P., Simmons, G.M., Jr., and Wharton, R.A., Jr., 1988, Climatological Observations and Predicted Sublimation Rates at Lake Hoare, Antarctica: Journal of Climate, v. 1, p. 715-728.

Conway, H., Hall, B.L., Denton, G.H., Gades, A.M., and Waddington, E.D., 1999, Past and Future Grounding-Line Retreat of the West Antarctic Ice Sheet: Science, v. 286, p. 280-283.

Cox, S.C., and Allibone, A.H., 1991, Petrogenesis of orthogneisses in the dry valleys region, South Victoria Land: Antarctic Science, v. 3 p. 405- 417.

Craig, J.R., Fortner, R.D., and Weand, B.L., 1974, Halite and Hydrohalite from Lake Bonney, Taylor Valley, Antarctica: Geology, v. 1, p. 389-390.

Cremer, H., Gore, D., Melles, M., and Roberts, D., 2003, Palaeoclimatic significance of late Quaternary diatom assemblages from southern Windmill Islands, East Antarctica: Palaeogeography, Palaeoclimatology, Palaeoecology, v. 135, p. 1-20.

Cunningham, W.L., Andrews, J.T., Jennings, A.E., Licht, K.J., and Leventer, A., 1999, Late Pleistocene-Holocene marine conditions in the Ross Sea, Antarctica: evidence from the diatom record: The Holocene, v. 9, p. 129-139.

Denton, G.H., Bockheim, J.G., Wilson, S.C., and Stuiver, M., 1989, Late Wisconsin and early Holocene glacial history, inner Ross embayment, Antarctica: Quaternary Research, v. 31, p. 151-182.

Denton, G.H., and Hughes, T.J., 2000, Reconstruction of the Ross Ice Drainage System, Antarctica, at the Last Glacial Maximum: Geografiska Annaler, Series A: Physical Geography, v. 82, p. 143-166.

Denton, G.H., and Marchant, D.R., 2000, The Geologic Basis for a Reconstruction of a Grounded Ice Sheet in McMurdo Sound, Antarctica, at the Last Glacial Maximum: Geografiska Annaler, Series A: Physical Geography, v. 82, p. 167-211.

Denton, G.H., Sugden, D.E., Marchant, D.R., Hall, B.L., and Wilch, T.I., 1993, East Antarctic Ice Sheet Sensitivity to Pliocene Climatic Change from a Dry Valleys Perspective: Geografiska Annaler. Series A. Physical Geography, v. 75, p. 155-204.

Domack, E.W., Jacobson, E.A., Shipp, S., and Anderson, J.B., 1999, Late Pleistocene-Holocene retreat of the West Antarctic Ice-Sheet system in the Ross Sea: Part 2 - Sedimentologic and stratigraphic signature: GSA Bulletin, v. 111, p. 1517-1536.

Doran, P.T., Berger, G.W., Lyons, W.B., Wharton Jr, R.A., Davisson, M.L., Southon, J., and Dibb, J.E., 1999, Dating Quaternary lacustrine sediments in the McMurdo Dry Valleys, Antarctica: Palaeogeography, Palaeoclimatology, Palaeoecology, v. 147, p. 223-239.

Doran, P.T., McKay, C.P., Clow, G.D., Dana, G.L., Fountain, A.G., Nylen, T., and Lyons, W.B., 2002a, Valley floor climate observations from the McMurdo Dry Valleys, Antarctica, 1986-2000: Journal of Geophysical Research, v. 107.

Doran, P.T., Priscu, J.C., Lyons, W.B., Walsh, J.E., Fountain, A.G., McKnight, D.M., Moorhead, D.L., Virginia, R.A., Wall, D.H., Clow, G.D., Fritsen, C.H., McKay, C.P., and Parsons, A.N., 2002b, Antarctic climate cooling and terrestrial ecosystem response: Nature, v. 415, p. 517-520.

Doran, P.T., Wharton Jr., R.A., and Lyons, W.B., 1994, Paleolimnology of the McMurdo Dry Valleys, Antarctica: Journal of Paleolimnology, v. 10, p. 85-114.

Dronkert, H., 1985, Evaporite models and sedimentology of Messinian and recent evaporites [PhD thesis]: Amsterdam, Universiteit van Amsterdam.

Ehrmann, W., and Polozek, K., 1999, The heavy mineral record in the Pliocene to Quaternary sediments of the CIROS-2 drill core, McMurdo Sound, Antarctica: Sedimentary Geology, v. 128, p. 223-244.

Elston, D.P., and Bressler, S.L., 1981, Magnetic stratigraphy of DVDP drill cores and late Cenozoic history of Taylor Valley, Transantarctic Mountains, Antarctica, in McGinnis, L.D., ed., Dry Valley Drilling Project, Volume 33: Antarctic Research Series: Washington, American Geophysical Union, p. 413-426.

Emslie, S.D., Coats, L., and Licht, K., 2007, A 45,000 yr record of Adelie penguins and climate change in the Ross Sea, Antarctica: Geology, v. 35, p. 61-64.

EPICA community members, 2004, Eight glacial cycles from an Antarctic ice core: Nature, v. 429, p. 623-628.

Eugster, H.P., 1980, Geochemistry of Evaporitic Lacustrine Deposits: Annual Review of Earth and Planetary Sciences, v. 8, p. 35-63.

Fountain, A.G., Dana, G.L., Lewis, K.J., Vaughn, B.H., and McKnight, D., 1998, Glaciers of the McMurdo Dry Valleys, Southern Victoria Land, Antarctica, in Priscu, J.C., ed., Ecosystem dynamics in a polar desert: The McMurdo Dry Valleys, Antarctica., Volume 72: Antarctic Research Series: Washington D.C., American Geophysical Union, p. 65-75.

Fountain, A.G., Lyons, W.B., Burkins, M.B., Dana, G.L., Doran, P.T., Lewis, K.J., McKnight, D.M., Moorhead, D.L., Parsons, A.N., Priscu, J.C., Wall, D.H., Wharton, R.A.J., and Virginia, R.A., 1999, Physical controls on the Taylor Valley ecosystem, Antarctica: BioScience, v. 49, p. 961-971.

Grootes, P.M., Nadeau, M.-J., and Rieck, A., 2004, ^{14}C-AMS at the Leibniz-Labor: radiometric dating and isotope research: Nuclear Instruments and Methods in Physics Research Section B: Beam Interactions with Materials and Atoms, v. 223-224, p. 55-61.

Grootes, P.M., Steig, E.J., Stuiver, M., Waddington, E.D., Morse, D.L., and Nadeau, M.-J., 2001, The Taylor Dome Antarctic ^{18}O Record and Globally Synchronous Changes in Climate: Quaternary Research, v. 56, p. 289-298.

Gunn, B., 2006a, Eastern Taylor Valley, image in Ross Sea Info – Landforms. URL: http://www.rosssea.info/pix/big/Antarc-Taylor-Valley-East.jpg (date of access: 8 Aug. 2007).

Gunn, B., 2006b, Taylor Glacier and Lake Bonney, image in Ross Sea Info – Geology. URL: http://www.rosssea.info/pix/big/Taylor_Gl-Lake_Bonney.jpg (date of access: 8 Aug. 2007).

Håkanson, L., and Jansson, M., 1983, Principles of Lake Sedimentology: Berlin, Springer-Verlag, 316 p.

Hall, B.L., 2003, Evidence for millennial-scale fluctuations of closed-basin lakes in the dry valleys, in GSA, ed., Geological Society of America - Annual Meeting, Volume 35, No. 6: Seattle, p. 464.

Hall, B.L., and Denton, G.H., 1999, New relative sea-level curves for the southern Scott Coast, Anarctica: evidence for Holocene deglaciation of the western Ross Sea: Journal of Quaternary Science, v. 14 p. 641-650.

Hall, B.L., and Denton, G.H., 2000a, Extent and Chronology of the Ross Sea Ice Sheet and the Wilson Piedmont Glacier along the Scott Coast at and Since the Last Glacial Maximum: Geografiska Annaler, Series A: Physical Geography, v. 82, p. 337-363.

Hall, B.L., and Denton, G.H., 2000b, Radiocarbon Chronology of Ross Sea Drift, Eastern Taylor Valley, Antarctica: Evidence for a Grounded Ice Sheet in the Ross Sea at the Last Glacial Maximum: Geografiska Annaler, Series A: Physical Geography, v. 82A, p. 305-336.

Hall, B.L., Denton, G.H., and Hendy, C.H., 2000, Evidence from Taylor Valley for a Grounded Ice Sheet in the Ross Sea, Antarctica: Geografiska Annaler, Series A: Physical Geography, v. 82, p. 275-303.

Hall, B.L., Denton, G.H., and Overturf, B., 2001, Glacial Lake Wright, a high-level Antarctic lake during the LGM and early Holocene: Antarctic Science, v. 13, p. 53-60.

Hall, B.L., Denton, G.H., Overturf, B., and Hendy, C.H., 2002, Glacial Lake Victoria, a high-level Antarctic Lake inferred from lacustrine deposits in Victoria Valley: Journal of Quaternary Science, v. 17, p. 697-706.

Hall, B.L., Hendy, C.H., and Denton, G.H., 2006a, Lake-ice conveyor deposits: Geomorphology, sedimentology, and importance in reconstructing the glacial history of the Dry Valleys: Geomorphology, v. 75, p. 143-156.

Hall, B.L., Hoelzel, A.R., Baroni, C., Denton, G.H., Le Boeuf, B.J., Overturf, B., and Topf, A.L., 2006b, Holocene elephant seal distribution implies warmer-than-present climate in the Ross Sea: Proceedings of the National Academy of Sciences, v. 103, p. 10213-10217.

Hendy, C.H., 2000a, Late Quaternary Lakes in the McMurdo Sound Region of Antarctica: Geografiska Annaler, Series A: Physical Geography, v. 82, p. 411-432.

Hendy, C.H., 2000b, The Role of Polar Lake Ice as a Filter for Glacial Lacustrine Sediments: Geografiska Annaler, Series A: Physical Geography, v. 82, p. 271-274.

Hendy, C.H., and Hall, B., 2006, The radiocarbon reservoir effect in proglacial lakes: Examples from Antarctica: Earth and Planetary Science Letters, v. 241, p. 413-421.

Hendy, C.H., Healy, T.R., Rayner, E.M., Shaw, J., and Wilson, A.T., 1979, Late Pleistocene Glacial Chronology of the Taylor Valley, Antarctica, and the Global Climate: Quaternary Research, v. 11, p. 172-184.

Hendy, C.H., Sadler, A.J., Denton, G.H., and Hall, B.L., 2000, Proglacial Lake-ice Conveyors: A New Mechanism for Deposition of Drift in Polar Environments: Geografiska Annaler, Series A: Physical Geography, v. 82, p. 249-270.

Hendy, C.H., Wilson, A.T., Popplewell, K.B., and House, D.A., 1977, Dating of geochemical events in Lake Bonney, Antarctica, and their relation to glacial and climate changes: New Zealand Journal of Geology & Geophysics, v. 20, p. 1103-1122.

Higgins, S.M., Denton, G.H., and Hendy, C.H., 2000a, Glacial Geomorphology of Bonney drift, Taylor Valley: Geografiska Annaler, v. 82 A, p. 365-390.

Higgins, S.M., Hendy, C.H., and Denton, G.H., 2000b, Geochronology of Bonney Drift, Taylor Valley, Antarctica: Evidence for Interglacial Expansions of Taylor Glacier: Geografiska Annaler, Series A: Physical Geography, v. 82, p. 391-409.

Hodgson, D.A., Doran, P.T., Roberts, D., and McMinn, A., 2004, Paleolimnological studies from the Antarctic and subantarctic islands, in Pienitz, R., Douglas, M.S.V., and Smol, J.P., eds., Long-term environmental change in Arctic and Antarctic lakes: Dordrecht, Springer, p. 419-474.

Hodgson, D.A., Noon, P.E., Vyverman, W., Bryant, C.L., Gore, D.B., Appleby, P., Gilmour, M., Verleyen, E., Sabbe, K., Ellis-Evans, J.C., and Wood, P.B., 2001, Were the Larsemann Hills ice-free through the Last Glacial Maximum?: Antarctic Science, v. 13, p. 440-454.

Hubbard, A., Lawson, W., Anderson, B., Hubbard, B.P., and Blatter, H., 2004, Evidence for subglacial ponding across Taylor Glacier, Dry Valleys, Antarctica: Annals of Glaciology, v. 39, p. 79-84.

Ingólfsson, Ó., 2004, Quaternary glacial and climate history of Antarctica, in Ehlers, J., and Gibbard, P.L., eds., Quaternary Glaciations - Extent and Chronology, Part III. , Elsevier, p. 3-43.

Kellogg, D.E., Stuiver, M., Kellogg, T.B., and Denton, G.H., 1980, Non-marine diatoms from late Wisconsin perched deltas in Taylor Valley, Antarctica: Palaeogeography, Palaeoclimatology, Palaeoecology, v. 30, p. 157-189.

Keys, J.R., and Williams, K., 1981, Origin of crystalline, cold desert salts in the McMurdo region, Antarctica: Geochimica et Cosmochimica Acta, v. 45, p. 2299-2309.

King, R.J., 2005, Minerals explained 42: Halite: Geology Today, v. 21, p. 153-157.

Krumgalz, B.S., 2001, Application of the Pitzer ion interaction model to natural hypersaline brines: Journal of Molecular Liquids, v. 91, p. 3-19.

Kulbe, T., Melles, M., Verkulich, S.R., and Pushina, Z.V., 2001, East Antarctic Climate and Environmental Variability over the Last 9400 Years Inferred from Marine Sediments of the Bunger Oasis: Arctic, Antarctic, and Alpine Research, v. 33, p. 223-230.

Kyle, P.R., 1981, Glacial history of McMurdo Sound area as indicated by the distribution andd nature of McMurdo Volcanic Group rocks, in McGinnis, L.D., ed., Dry Valley Drilling Project, Volume 33: Antarctic Research Series: Washington, American Geophysical Union, p. 403-412.

Last, W.M., 1993a, Geolimnology of Freefight Lake: an unusual hypersaline lake in the northern Great Plains of western Canada: Sedimentology, v. 40, p. 431-448.

Last, W.M., 1993b, Rates of sediment deposition in a hypersaline lake in the northern Great Plains of western Canada: International Journal of Salt Lake Research, v. 2, p. 47-58.

Lawrence, M.J.F., and Hendy, C.H., 1985, Water column and sediment characteristics of Lake Fryxell, Taylor Valley, Antarctica: New Zealand Journal of Geology and Geophysics, v. 28, p. 543-552.

Lawrence, M.J.F., and Hendy, C.H., 1989, Carbonate deposition and Ross Sea ice advance, Fryxell basin, Taylor Valley, Antarctica: New Zealand Journal of Geology and Geophysics, v. 32, p. 267-277.

Lawson, J., Doran, P.T., Kenig, F., des Marais, D.J., and Priscu, J.C., 2004, Stable Carbon and Nitrogen Isotopic Composition of Benthic and Pelagic Organic Matter in Lakes of the McMurdo Dry Valleys, Antarctica: Aquatic Geochemistry, v. 10, p. 269-301.

Lawson Knoepfle, J., Doran, P.T., Kenig, F., and Lyons, W.B., in prep., Radiocarbon abundance and reservoir effects in lakes of the McMurdo Dry Valleys, Antarctica.

Licht, K.J., Dunbar, N.W., Andrews, J.T., and Jennings, A.E., 1999, Distinguishing subglacial till and glacial marine diamictons in the western Ross Sea, Antarctica; implications for a last glacial maximum grounding line: Geol Soc Am Bull, v. 111, p. 91-103.

Licht, K.J., Jennings, A.E., Andrews, J.T., and Williams, K.M., 1996, Chronology of late Wisconsin ice retreat from the western Ross Sea, Antarctica: Geology, v. 24, p. 223-226.

Lowenstein, T.K., and Hardie, L.A., 1985, Criteria for the recognition of salt-pan evaporites: Sedimentology, v. 32, p. 627-644.

Lyons, W.B., Fountain, A.G., Doran, P., Priscu, J.C., Neumann, K., and Welch, K.A., 2000, Importance of landscape position and legacy: the evolution of the lakes in Taylor Valley, Antarctica: Freshwater Biology, v. 43, p. 355-367.

Lyons, W.B., Frape, S.K., and Welch, K.A., 1999, History of McMurdo Dry Valley lakes, Antarctica, from stable chlorine isotope data: Geology, v. 27, p. 527-530.

Lyons, W.B., Mayewski, P.A., Donahue, P., and Cassidy, D., 1985, A preliminary study of the sedimentary history of Lake Vanda, Antarctica: climatic implications: New Zealand Journal of Marine and Freshwater Research, v. 19, p. 253-260.

Lyons, W.B., and Priscu, J., n.d., McMurdo Dry Valley Lake Descriptions. McMurdo Dry Valleys. LTER Data Bank: knb-lter-mcm.0038.2 [Database]. URL: http://www.mcmlter.org/data/lakes/locations/lakedsc.dat (date of access: 4 April 2008).

Lyons, W.B., Tyler, S.W., Wharton, R.A., McKnight, D.M., and Vaughn, B.H., 1998, A Late Holocene desiccation of Lake Hoare and Lake Fryxell, McMurdo Dry Valleys, Antarctica: Antarctic Science, v. 10, p. 247-256.

Lyons, W.B., and Welch, K., 2007, Limnological Chemistry / Ion Concentrations and Si. McMurdo Dry Valleys LTER Data Bank: knb-lter-mcm.0062.4 [Database] URL: http://www.mcmlter.org/queries/limno_results.jsp?begDate=01/01/0001&endDate=01/01/3000&dataType=CHEMISTRY (date of access: 4 April 2008).

Lyons, W.B., Welch, K.A., Snyder, G., Olesik, J., Graham, E.Y., Marion, G.M., and Poreda, R.J., 2005, Halogen geochemistry of the McMurdo dry valleys lakes, Antarctica: Clues to the origin of solutes and lake evolution: Geochimica et Cosmochimica Acta, v. 69, p. 305-323.

Masson, V., Vimeux, F., Jouzel, J., Morgan, V., Delmotte, M., Ciais, P., Hammer, C., Johnsen, S., Lipenkov, V.Y., and Mosley-Thompson, E., 2000, Holocene Climate Variability in Antarctica Based on 11 Ice-Core Isotopic Records: Quaternary Research, v. 54, p. 348-358.

Matsubaya, O., Sakai, H., Torii, T., Burton, H., and Kerry, K., 1979, Antarctic saline lakes--stable isotopic ratios, chemical compositions and evolution: Geochimica et Cosmochimica Acta, v. 43, p. 7-25.

McKay, R.M., Dunbar, G.B., Naish, T.R., Barrett, P.J., Carter, L., and Harper, M., 2008, Retreat history of the Ross Ice Sheet (Shelf) since the Last Glacial Maximum from deep-basin sediment cores around Ross Island: Palaeogeography, Palaeoclimatology, Palaeoecology, v. 260, p. 245-261.

McKnight, D.M., and Andrews, E.D., 1993, Potential hydrologic and geochemical consequences of the 1992 merging of Lake Chad with Lake Hoare in Taylor Valley.: Antarctic Journal of the United States, v. 28, p. 249-251.

McKnight, D.M., Niyogi, D.K., Alger, A.S., Bomblies, A., Conovitz, P.A., and Tate, C.M., 1999, Dry Valley Streams in Antarctica: Ecosystems Waiting for Water: BioScience, v. 49, p. 985-995.

Mikucki, J.A., Foreman, C.M., Sattler, B., Berry Lyons, W., and Priscu, J.C., 2004, Geomicrobiology of Blood Falls: An Iron-Rich Saline Discharge at the Terminus of the Taylor Glacier, Antarctica: Aquatic Geochemistry, v. 10, p. 199-220.

Nedell, S.S., Andersen, D.W., Squyres, S.W., and Love, F.G., 1987, Sedimentation in ice-covered Lake Hoare, Antarctica: Sedimentology, v. 34, p. 1093-1106.

Neumann, K., Lyons, W.B., Priscu, J.C., Desmarais, D.J., and Welch, K.A., 2004, The carbon isotopic composition of dissolved inorganic carbon in perennially ice-covered Antarctic lakes: searching for a biogenic signature: Annals of Glaciology, v. 39, p. 518-524.

Neumann, M., and Ehrmann, W., 2001, Mineralogy of sediments from CRP-3, Victoria Land Basin, Antarctica, as revealed by X-ray diffraction: Terra Antartica v. 8, p. 523-532.

Petit, J.R., Jouzel, J., Raynaud, D., Barkov, N.I., Barnola, J.M., Basile, I., Bender, M., Chappellaz, J., Davis, M., Delaygue, G., Delmotte, M., Kotlyakov, V.M., Legrand, M., Lipenkov, V.Y., Lorius, C., Pepin, L., Ritz, C., Saltzman, E., and Stievenard, M., 1999, Climate and atmospheric history of the past 420,000 years from the Vostok ice core, Antarctica: Nature, v. 399, p. 429-436.

Petschick, R., 2001, MacDiff 4.2.5 Freeware Scientific Graphical Analysis of X-Ray Diffraction Profiles: Frankfurt/Main.

Poreda, R.J., Hunt, A.G., Lyons, W.B., and Welch, K.A., 2004, The Helium Isotopic Chemistry of Lake Bonney, Taylor Valley, Antarctica: Timing of Late Holocene Climate Change in Antarctica: Aquatic Geochemistry, v. 10, p. 353-371.

Porter, S.C., and Beget, J.E., 1981, Provenance and depositional environments of Late Cenozoic sediments in Permafrost cores from lower Taylor Valley, Antarctica, in McGinnis, L.D., ed., Dry Valley Drilling Project, Volume 33: Antarctic Research Series: Washington, American Geophysical Union, p. 351-363.

Powell, R.D., 1981, Sedimentation conditions in Taylor Valley, Antarctica, inferred from textural analysis of DVDP cores, in McGinnis, L.D., ed., Dry Valley Drilling Project, Volume 33: Antarctic Research Series: Washington, American Geophysical Union, p. 331-349.

Priscu, J.C., 1998, Ecosystem dynamics in a polar desert: The McMurdo Dry Valleys, Antarctica.: Washington D.C., American Geophysical Union, 369 p.

SCAR (Scientific Committee on Antarctic Research), 2007, Some Antarctic Statistics. URL: http://www.scar.org/information/statistics/index.html (date of access: 24 Oct. 2008).

Schmok, J.P., and Waddington, B.S., 1996, Lakes Hoare, Fryxell and Bonney: Geophysical Determination of Bathymetry and Morphometry: Burnaby, British Columbia, Canada, Golder Associates Ltd.

Schreiber, B.C., and El Tabakh, M., 2000, Deposition and early alteration of evaporites: Sedimentology, v. 47, p. 215-238.

Schwab, M., 1998, Rekonstruktion der spätquartären Klima- und Umweltgeschichte der Schirmacheroase und des Wohlthatmassivs (Ostantarktika) [PhD thesis]: Potsdam, University Potsdam.

Shipp, S., Anderson, J., and Domack, E.W., 1999, Late Pleistocene–Holocene retreat of the West Antarctic Ice-Sheet system in the Ross Sea: Part 1 - Geophysical results.: GSA Bulletin, v. 111, p. 1486-1516.

Smillie, R.W., 1992, Suite subdivision and petrological evolution of granitoids from the Taylor Valley and Ferrar Glacier region, south Victoria Land: Antarctic Science, v. 4 p. 71-87.

Smith, G.I., and Friedman, I., 1993, Lithology and paleoclimatic implications of lacustrine deposits around Lake Vanda and Don Juan Pond, Antarctica: Antarctic Research Series, v. 59, p. 83-94.

Spaulding, S.A., McKnight, D.M., Stoermer, E.F., and Doran, P.T., 1997, Diatoms in sediments of perennially ice-covered Lake Hoare, and implications for interpreting lake history in the McMurdo Dry Valleys of Antarctica: Journal of Paleolimnology, v. 17, p. 403-420.

Spigel, R.H., and Priscu, J.C., 1996, Evolution of temperature and salt structure of Lake Bonney, a chemically stratified Antarctic lake: Hydrobiologia, v. 321, p. 177-190.

Squyres, S.W., Andersen, D.W., Nedell, S.S., and Wharton, R.A., 1991, Lake Hoare, Antarctica: sedimentation through a thick perennial ice cover: Sedimentology, v. 38, p. 363-379.

Steig, E.J., Brook, E.J., White, J.W., nbsp, C, Sucher, C.M., Bender, M.L., Lehman, S.J., Morse, D.L., Waddington, E.D., and Clow, G.D., 1998, Synchronous Climate Changes in Antarctica and the North Atlantic: Science, v. 282, p. 92-95.

Steig, E.J., Morse, D.L., Waddington, E.D., Stuiver, M., Grootes, P.M., Mayewski, P.A., Twickler, M.S., and Whitlow, S.I., 2000, Wisconsinan and Holocene Climate History from an Ice Core at Taylor Dome, Western Ross Embayment, Antarctica: Geografiska Annaler, Series A: Physical Geography, v. 82, p. 213-235.

Stuiver, M., Denton, G.H., Hughes, T.J., and Fastook, J.L., 1981, History of the Marine Ice Sheet in West Antarctica during the Last Glaciation: A Working Hypothesis, in Denton, G.H., and Hughes, T.J., eds., The Last Great Ice Sheets: New York, Chichester, Brisbane, Toronto, Wiley Interscience Publication, p. 319-436.

Tyler, S.W., Cook, P.G., Butt, A.Z., Thomas, J.M., Doran, P.T., and Lyons, W.B., 1998, Evidence of deep circulation in two perennially ice-covered Antarktik lakes: Limnology and Oceanography, v. 43 p. 625-635.

U.S. Geological Survey, 2006, Satellite image Map of Antarctica, AVHRR Mosaic Color Composite. URL: http://Terraweb.wr.usgs.gov/TRS/projects/Antarctica/AVHRR.html (date of access: 3 May 2006).

U.S. Geological Survey, 2007, Landsat Image Mosaic Antarctica (LIMA) – McMurdo Dry Valleys. URL: http://lima.usgs.gov/index.php (date of access: 10 June 2007).

Verleyen, E., Hodgson, D.A., Sabbe, K., and Vyverman, W., 2004, Late Quaternary deglaciation and climate history of the Larsemann Hills (East Antarctica): Journal of Quaternary Science, v. 19, p. 361-375.

Wagner, B., 2003, The Expeditions Amery Oasis, East Antarctica, in 2001/02 and Taylor Valley, Southern Victoria Land, in 2002: Reports on Polar and Marine Research, v. 460, p. 1-69.

Wagner, B., Cremer, H., Hultzsch, N., Gore, D.B., and Melles, M., 2004, Late Pleistocene and Holocene history of Lake Terrasovoje, Amery Oasis, East Antarctica, and its climatic and environmental implications: Journal of Paleolimnology, v. 32, p. 321-339.

Wagner, B., Melles, M., Doran, P.T., Kenig, F., Forman, S.L., Pierau, R., and Allen, P., 2006, Glacial and postglacial sedimentation in the Fryxell basin, Taylor Valley, southern Victoria Land, Antarctica: Palaeogeography, Palaeoclimatology, Palaeoecology, v. 241, p. 320-337.

Wharton, R.A., Mackay, A.W., Simmons, G.M., and Parker, B.C., 1986, Oxygen budget of perennially ice-covered Antarctic lake: Limnology and Oceanography, v. 31 p. 437-443.

Wharton, R.A.J., McKay, C.P., Clow, G.D., Andersen, D.T., Simmons, G.M.J., and Love, F.G., 1992, Changes in thickness and lake level of Lake Hoare, Antarctica: Implications for local climatic change: Journal of Geophysical Research, v. 97, p. 3503-3513.

References

Wharton, R.A.J., Parker, B.C., and Simmons, G.M.J., 1983, Distribution, species composition and morphology of algalmats in Antarctic dry valley lakes: Phycologia, v. 22, p. 355-365.

Wharton, R.A.J., Parker, B.C., Simmons, G.M.J., Seaburg, K.G., and Love, F.G., 1982, Biogenic calcite structures forming in Lake Fryxell, Antarctica: Nature, v. 295, p. 403-405.

Wharton, R.A.J., Simmons, G.M.J., and McKay, C.P., 1989, Perennially ice-covered Lake Hoare, Antarctica: physical environment, biology and sedimentation: Hydrobiologia, v. 172, p. 305-320.

Whittaker, T.E., Hall, B.L., Hendy, C.H., and Spaulding, S.A., 2008, Holocene depositional environments and surface-level changes at Lake Fryxell, Antarctica: The Holocene, v. 18, p. 775-786.

Wilson, A.T., Hendy, C.H., Healy, T.R., Gumbley, J.W., Field, A.B., and Reynolds, C.P., 1974, Dry valley lake sediments: a record of Cenozoic climatic events: Antarctic Journal of the U.S., v. 9, p. 134-135.

Die VDM Verlagsservicegesellschaft sucht für wissenschaftliche Verlage abgeschlossene und herausragende

Dissertationen, Habilitationen, Diplomarbeiten, Master Theses, Magisterarbeiten usw.

für die kostenlose Publikation als Fachbuch.

Sie verfügen über eine Arbeit, die hohen inhaltlichen und formalen Ansprüchen genügt, und haben Interesse an einer honorarvergüteten Publikation?

Dann senden Sie bitte erste Informationen über sich und Ihre Arbeit per Email an *info@vdm-vsg.de*.

Sie erhalten kurzfristig unser Feedback!

VDM Verlagsservicegesellschaft mbH
Dudweiler Landstr. 99   Telefon +49 681 3720 174
D - 66123 Saarbrücken   Fax   +49 681 3720 1749
www.vdm-vsg.de

Die VDM Verlagsservicegesellschaft mbH vertritt

Printed by Books on Demand GmbH, Norderstedt / Germany